O9-CFU-692

Outcasts from Evolution

SCIENTIFIC ATTITUDES OF RACIAL INFERIORITY, 1859–1900

John S. Haller, Jr.

SOUTHERN ILLINOIS UNIVERSITY PRESS
Carbondale & Edwardsville

Permission has been granted by the following to reprint my material which appears in different form in several chapters of this book: *Journal of the History of Medicine and Allied Sciences*, for "Concepts of Race Inferiority in Nineteenth-Century Anthropology," XXV (Jan., 1970), and "Race, Mortality, and Life Insurance: Negro Vital Statistics in the Late Nineteenth Century," XXV (July, 1970); The Johns Hopkins University Press, *Bulletin of the History of Medicine*, for "The Physician versus the Negro: Medical and Anthropological Concepts of Race in the Late Nineteenth Century," XLIV (Mar.–Apr., 1970); American Anthropological Association, *American Anthropologist*, for "The Species Problem: Nineteenth-Century Concepts of Racial Inferiority in the Origin of Man Controversy," LXXII (Dec., 1970), and "Race and the Concept of 'Progress' in Nineteenth-Century American Ethnology," LXXIII (June, 1971); Peabody Essex Museum, Salem, Massachusetts, *Essex Institute Historical Collections*, for "Nathaniel Southgate Shaler: A Portrait of Nineteenth-Century Academic Thinking on Race," CVII (Apr., 1971); and *Civil War History*, for "Civil War Anthropometry: The Making of a Racial Ideology," XVI (Dec., 1970).

Published by agreement with the University of Illinois Press

Printed in the United States of America
Production supervised by Duane Perkins
00 99 98 97 5 4 3 2

Library of Congress Cataloging-in-Publication Data

Haller, John S.
Outcasts from evolution : scientific attitudes of racial inferiority, 1859–1900 / John S. Haller, Jr.
p. cm.
Originally published: Urbana : University of Illinois Press, [1971]. With new pref.
Includes bibliographical references (p.) and index.
1. Race. 2. Black race. 3. United States—Race relations. 4. Anthropology—History. I. Title.
GN269.H34 1995
305.8—dc20 94-48231
ISBN 0-8093-1982-9 (paperback) CIP

Cover illustration: From Ranson Dexter, "The Facial Angle," *Popular Science Monthly* (1874). Courtesy of Indiana University.

The paper used in this publication meets the minimum requirements of American National Standard for Information Sciences—Permanence of Paper for Printed Library Materials, ANSI Z39.48-1984. ♾

TO MY PARENTS & ROBIN

Contents

Illustrations

Preface to New Edition

NEARLY TWENTY-FIVE YEARS have passed since the original publication of this book. That it has stood the test of time lies in large part, I believe, because I chose not to rewrite the past to serve the present of 1971. This singular point merits restatement since our ability to understand and appreciate the past lies in our willingness to accept it for what it is, in itself, and not for what we might select from it to serve later generations. In saying this I do not mean to discount whatever meaning the past may have for the present but only that any revisionism which distorts or otherwise misrepresents the past's own values is not history.

The period from Darwin to Mendel was one whose theories on race, both "liberal" and "conservative," accepted only variations on the implicit assumption of racial inferiority. Those who argued equality of the races—either biological or legal—effectively lost credence within the reigning science and social science paradigm. This paradigm not only determined the relative value of the races but also helped to delineate social categories and justify the century's efforts at social engineering. The results were self-evident. The subject of race inferiority was simply beyond critical reach in the late nineteenth century. "Society," wrote Henry Adams, "offered the profile of a long, straggling caravan, stretching loosely toward the prairies, its few score of leaders far in advance and its millions of immigrants, negroes, and Indians far in the rear, somewhere in archaic time." This was not the solipsistic invention of a few; rather it was the mythology of a nation. Its dark implications, which have since grated upon the sensibilities of our own age, helped nonetheless to define individual rights and liberties, articulate the ideals of a good society, reveal the true vigor of popular sovereignty, and make the world safe for democracy. It is not surprising, then, that when we try to appraise this period in terms meaningful to the twentieth century, it becomes elusive and evanescent. Such is the paradox of our history.

John S. Haller, Jr.
March, 1995

Preface

MANY HISTORIANS, caught up in the belief that racial harmony was always part of America's liberal tradition, have dealt with the subject of race relations by trying to discover our society's best values within the context of the democratic beliefs of the past. But quite often their attempts to reinforce their own optimistic hopes for racial harmony with the aid of the democratic ideology of the Enlightenment, of Jacksonianism, of the Transcendentalists, and of the post–Civil War generation of reformers have been achieved only by losing some of the spirit of the ages studied. In justifying themselves to posterity and in attempting to preserve the dignity of the men who laid the foundations of their own liberalism, many historians have ended in the futile gesture of teaching a moral while at the same time losing the historical content of the past. The past, however, contains its own philosophy, and those historians who look to the nineteenth century for the verification of an egalitarian viewpoint in race relations tend to distort immeasurably the philosophic framework of that century.

I have tried, within the limitations that are always present in a work of this nature, to see the nineteenth century's attitudes of race within its own framework—a framework which was much closer to the seventeenth- and eighteenth-century concepts of man than to the twentieth century's search for an egalitarian society. It was a century whose racial theories, both "liberal" and "conservative," tended to perpetuate an enduring image masked with assorted variations on the single theme of permanent racial inferiority.

This book is a study of the currents of intellectual thought from 1859 to 1900, centering on the development of America's scientific attitudes of race. Marked at one end by the publication of Darwin's *Origin of Species* and at the other by the rediscovery of Mendel's law of inheritance in 1900, this work seeks to re-create an internal portrait of anthropology and the application of its ideas in medicine, psychology, ethnology, and sociology during the so-called "heroic" age of evolutionary-minded synthesizers. The period was important in the development of science in America, for scientific ideas quickly entered the popular culture through the spirited efforts of men like Herbert Spencer, John Fiske, John Wesley Powell, Edward D. Cope, Frederick Hoffman, Joseph LeConte, Nathaniel Shaler, and others who sought to acquaint society with the "truths" of evolution and the new evolutionary methodology and to apply those "truths" to the study of man. Many of the men who formulated the period's intellectual ideas not only helped to justify the "radical" Jeffersonianism of Reconstruction politics but willingly contributed to the disintegration of those ideals in the efforts during the later decades of the century to isolate the Negro through Jim Crow laws and political disfranchisement.

What was at once the worst of nineteenth-century America in the sense that we now judge its racial attitudes was also, ironically, the best that American culture had to offer. The sciences, those "portions of human knowledge [that] have been more or less generalized, systemized, and verified," became the means through which both scientists and social scientists sought to determine the relative value of the races of man, delineate social categories, and even justify the rationale of race legislation. The majority were Spencerian social Darwinists or neo-Lamarckians, whose attitudes concerning the possibilities of Caucasian race progress through rapid evolutionary improvement by means of the inheritance of acquired characteristics were optimistic. Unlike the neo-Darwinists Francis Galton and August Weismann, whose conservative hereditarian approach to race character virtually denied the possibility of modification through life experiences, the American environmentalists accepted a use-inheritance explanation of Caucasian race character. Mirroring the optimism of nineteenth-century humanitarian reform movements, the behavioral sciences in America encouraged modification through man-made social activity. Ironically, however, the environmentalist tradition became weighted with hereditarian ideas as soon as race analysis focused upon the non-Aryan peoples. Nineteenth-century assumptions of racial inferiority precluded a firm commitment to the agency of use-inheritance. The humanitarian and universalist implications of the environmentalist tradition existed on the horns of a democratic dilemma—America had first to be made "safe" before it could become democratic. Those elements which, according to scientists and social scientists, were unregenerate in terms both of physiological status and of America's exclusive political and moral mission were denied assimilation into American society. Believing that failures in earlier stages of evolution had limited brain size and quality of the lower races, these scientists and social scientists suggested that the environment no longer operated as strongly in the present as it once had in the past. Evolution had already come to an end among the lower races, making them unfit for future race development. While the Caucasian maintained an active role in modifying the environment, the lower races broke into the modern world as mere "survivals" from the past, mentally incapable of shouldering the burdens of complex civilization and slowly deteriorating structurally to a point when, at some time in the future, they would become extinct, thus ultimately solving the race problem.

Most of the environmentalists were not outspoken racists. As leading physicians, anthropologists, educators, paleontologists, and sociologists, their views on race inferiority, at once assumed and "proven" within the context of their framework, were not the primary subject of their concern but, rather, were elements which partially formed the foundation of their larger intellectualizations. Most of these men were recognized for ideas other than those they expressed on race. For this reason, the very articulation of their ideas is crucial to any attempt to dissect the late nineteenth-century's attitudes of race. Their ideas stand much closer to the nation's racial ideology than those examples often cited which, irrespective of publication date, are obvious almost by the very vindictiveness of their titles. The ideas presented by these men are much more clinical (though not necessarily less vindictive), even "scientific," tucked away in works of more mammoth importance.

What this study intends to show is the manner in which their science provided a vocabulary and a set of concepts which rationalized and helped to justify the value system upon which the idea of racial inferiority rested in American thought. For the intellectual of this period, the life history of America sharply divided between the Caucasian and the "colored" races—an almost commonplace division made by both "liberals" and "conservatives" of the country. The men studied in this book, through their ideas as well as their disciples, bridged the western culture, and as their scientific beliefs became household words in the circles of academia, so, too, their attitudes of race became much more significant in the context of their larger renown. For many educated Americans who shunned the stigma of racial prejudice, science became an instrument which "verified" the presumptive inferiority of the Negro and rationalized the politics of disfranchisement and segregation into a social-scientific terminology that satisfied the troubled conscience of the middle class. To understand attitudes of racial inferiority in the context of nineteenth-century science and social science is a first step in fathoming the depth of race prejudice in our own day. Inferiority was at the very foundation of their evolutionary framework and, remaining there, rose to the pinnacle of "truth" with the myth of scientific certainty. To see racial prejudices in their scientific robes is to understand why, despite later conceptual changes in evolution and methodology, attitudes of racial inferiority have continued to plague western culture.

I wish to acknowledge my special gratitude and indebtedness to George H. Callcott for his constant encouragement, unflagging interest, and friendly counsel. It was he who made the original suggestion out of which this study grew. His indispensable guidance as well as his untiring aid have made me deeply grateful. I am particularly thankful to Francis C. Haber, who helped to readjust the level of my intellectual sights. His suggestive criticisms have shown me the limitations under which all of us work who deal in the slightest way with science. I am also indebted to the generous encouragement, analysis, and criticism of Idus Newby. His valuable suggestions and perceptive reading saved me from an assortment of problems. Those that remain are my own responsibility. My special thanks go also to Dorothy T. Hanks, K. Janelle Wilson, and Lucy Keister of the National Library of Medicine, Pat Havalice, Larry Fortado, and Agota Kuperman, research librarians at Indiana University, Harry G. Day, associate dean of Research and Advanced Studies at Indiana University, and Rosalie Zak, my patient and untiring secretary. In addition, my intellectual debts go to James Flack, Renny McLeod, and Roger Daday, my colleagues Neil Betten and Raymond Mohl, and the nameless students whose questions helped to clarify my ideas. And to my wife Robin, I am indebted for her patient reading, gentle criticism, and clerical assistance which added immeasurably to the finished product. The book would not have been possible without her. I also appreciate the courtesies extended by the National Library of Medicine, John Crerar Library, Midwest Center for Research Studies, Library of Congress, American Philosophical Society, Vanderbilt University Medical School Library, Billings Medical Library at the University of Chicago, Frances Carrick Thomas Library of Transylvania College, New York Academy of Medicine Library, University of California Medical Center, Rudolph Matas Medical Library of Tulane University, College of Physicians of

Philadelphia Library, Bio-Medical Library of the University of Minnesota, Enoch Pratt Free Library of Baltimore, and the Medical Center Library of the University of Michigan.

John S. Haller, Jr.
September, 1970

Outcasts from Evolution

I *Attitudes of Racial Inferiority in Nineteenth-Century Anthropometry*

LATE NINETEENTH-CENTURY ANTHROPOLOGY inherited a problem that had been building up with ever-increasing intensity since the voyages of discovery and exploration. As travelers penetrated the various nonwestern cultures of the world, their descriptive accounts of the variations among groups of men multiplied by the hundreds. Classification of these groups into varieties or races of men was attempted by eighteenth- and early nineteenth-century naturalists, but they were unable to formulate a common index to distinguish one race of men from another. To visually identify differences was one thing, but to determine a method for measurement and an index for tracing affinities among the various races was a far more vexatious undertaking. For the nineteenth-century anthropologist, anthropometry, or anatomical measurement, became a focal point in the study of man.[1]

[1] Alfred R. Hall, *The Scientific Revolution, 1500–1800* (New York, 1954), 283; J. Barnard Davis, "Measurements as a Means of Distinguishing Races," *American*

Carl von Linnaeus (1707–1778), who developed a taxonomic system based on a criterion of skin color, laid the basis for nineteenth-century racial classification. Linnaeus properly began the science of anthropology. Although color classification of races dated back to the ancient Egyptians, anthropologists referred to Linnaeus's taxonomy in his *Systema naturae* (1735) as the first modern study of man. While Linnaeus advanced classification with his use of a color criterion, he also fixed on his four families of man certain moral and intellectual peculiarities that continued into the nineteenth-century anthropological vocabulary. He described *Homo Americanus* as reddish, choleric, obstinate, contented, and regulated by customs; *Homo Europaeus* as white, fickle, sanguine, blue-eyed, gentle, and governed by laws; *Homo Asiaticus* as sallow, grave, dignified, avaricious, and ruled by opinions; and *Homo Afer* as black, phlegmatic, cunning, lazy, lustful, careless, and governed by caprice. These "insights" into what Linnaeus divined as racial character, personality traits, behavior, intelligence, language, and a host of other related categories were transmitted into subsequent attempts at a science of classification and became more fixed than the races themselves.[2]

Johann Friedrich Blumenbach (1752–1840), a professor at Göttingen, designated five races or varieties of man in the second edition of his treatise *On the Natural Variety of Mankind* (1781). His division into Caucasian, Mongolian, American, Ethiopian, and Malayan races, with the added Linnaean descriptive peculiarities, became the subsequent basis of most nineteenth-century anthropometrical studies. While Linnaeus founded his classificatory

---

Journal of Science, LXXIX (1860), 329; Philip D. Curtin, *The Image of Africa: British Ideas and Action, 1780–1850* (Madison, Wis., 1964); Katherine George, "The Civilized West Looks at Primitive Africa: 1400–1800," *Isis*, XLIX (1958), 62–72.

[2] Carl von Linnaeus, *A General System of Nature*, 7 vols. (London, 1806), I, 9; Louis L. Snyder, *Race: A History of Modern Ethnic Theories* (New York, 1939), 12; Josiah C. Nott and George R. Gliddon, *Types of Mankind* (Philadelphia, 1854), 84–86, 250–52; Alfred C. Haddon, *History of Anthropology* (London, 1910), 7–11; John C. Greene, "Some Early Speculations on the Origin of Human Races," *American Anthropologist*, n.s., LVI (Feb., 1954), 31–41; Walter Scheidt, "The Concepts of Race in Anthropology and the Divisions into Human Races from Linnaeus to Deniker," in Earl W. Count, ed., *This Is Race* (New York, 1950), 354–91.

system principally upon skin color, Blumenbach considered a combination of color, hair, skull, and facial characteristics as fundamental means for classifying the five varieties of man. Central to his study was the Caucasian, a term which he originated. He took the name "Caucasian" from Mount Caucasus because its southern slope had cradled what he felt to be the most beautiful race of men, the Georgian. The Caucasus, near Mount Ararat, upon which the biblical Ark came to rest after the Flood, seemed the appropriate source for the original race of man.

> For, in the first place, the stock displays, as we have seen, the most beautiful form of the skull, from which, as from a mean and primeval type, the others diverge by most easy gradations on both sides to the two ultimate extremes (that is, on the one side the Mongolian, on the other the Ethiopian). Besides, it is white in color, which we may fairly assume to have been the primitive color of mankind, since, as we have shown above, it is very easy for that to degenerate into brown, but very much more difficult for dark to become white, when the secretion and precipitation of this carbonaceous pigment has once deeply struck root.[3]

From the beginning Blumenbach accepted an unequal zoological importance in the five racial varieties. He looked upon the Caucasian, Mongolian, and Ethiopian as the three principal races. The Caucasian was not only the most beautiful of the varieties but also the basis from which the others derived. Using the shape of the skull as his criterion, he argued that the Mongolian and Ethiopian were extreme degenerations from the original autochthon. He relegated the two other races, American and Malayan, to transitional phases of only minor importance. The American represented the transitional passage from Caucasian to Mongolian, while the Malayan was the intermediate variety in the passage from Caucasian to Ethiopian.[4]

Generally, anthropologists accepted Blumenbach's categories.

[3] Thomas Bendyshe, ed., *The Anthropological Treatises of Johann Friedrich Blumenbach* (London, 1865), 269; Paul Topinard, *Anthropology* (London, 1878), 199.

[4] Bendyshe draws these inferences from Blumenbach's writings (*Anthropological Treatises*, x–xi).

Successive students of classification, men like William Lawrence (1783–1867), James Cowles Prichard (1786–1848), and Theodore Waitz (1821–1864), took for granted the reality of the five races. Blumenbach's theory was put forth at a time when the origination of man was believed to have been an act of special creation. Even those who later accepted Darwin's theory of evolution were inclined to retain Blumenbach's divisions. This they did by defining the races as "forms" that diverged from the proto-stock at some early stage in life and which remained separate and distinct for long periods of time to the point of becoming "fixed" in their characteristics. In fact, most nineteenth-century anthropometrical researches generally accepted the divisions which Blumenbach made of the varieties of man and incorporated his divisions into their own schemata with little hesitation. Rather than dispute the division, they moved on to develop measuring devices which gave additional accent to the previously established divisions. Furthermore, their instruments established stock differences within each of the major divisions and went on to prove a gradation from the anthropoid through the varieties of man.

Although many of the late nineteenth-century anthropologists built upon Blumenbach's five divisions of man, they distorted much of his scientific framework by elaborating on the close relationship between the Negro and the orangoutan. In Blumenbach's opinion Linnaeus had made a fundamental mistake in placing man in the animal kingdom. To infer the close resemblance of the Negro to the orangoutan distorted the basic unity of man and his spiritual and moral integrity. In spite of the African's many misfortunes, Blumenbach "reckon[ed] it among the most humane and the bravest men; authors, learned men and poets." Indeed, according to Blumenbach, "all men are born, or might have been born from the same man." The Negroes were "our black brothers."[5]

From an initial emphasis upon enumeration of the races, the nineteenth-century science of man moved on into anthropometry

[5] M. Flourens, "Memoir of Blumenbach," *ibid.*, 57, 60; Blumenbach, "Observations on the Bodily Conformation and Mental Capacity of the Negroes," *Philosophical Magazine*, III (1799), 141–47.

through ingenious efforts to determine racial peculiarities. The hallmark of anthropology in the nineteenth century was anthropometry.[6] From Paul Broca in France to Herbert Spencer in England, from Joseph Henry of the Smithsonian Institution to the diligent statisticians of the American Union Army, anthropologists used anthropometry to distinguish and clarify racial differences, both among the five major varieties designated by Blumenbach and among the various "stocks" within particular race groups. Anthropometrists in Europe and America experimented with measurements of skull shape, hair pile, skin color, temperament, and political belief in order to determine the reality and ranking of dozens or hundreds of stocks.[7]

Some anthropometrists looked to hair as a means of classification. "Human hairs differ in the manner in which they are imbedded and by the differing forms of the transverse sections when cut open, straight hair being circular and wavy hair being elliptical in section." The Frenchman Bory de Saint-Vincent distinguished between straight-haired and woolly-haired species of man, a division that corresponded to Joseph Julien Virey's two species: black and white. According to the Frenchman Franz Pruner-Bey, hair was a legitimate basis for racial classification: (1) flat or woolly hair of Negroes, (2) large and coarse cylindrical hair of Mongols, Chinese, Malays, and Americans, and (3) "hair intermediate in size and shape" of Europeans.[8] In the United States, however, hair classification never really became a legitimately recognized means of scientific racial classification. Too many "unscientific" enthusiasts distorted anatomical differences by going beyond classification itself.

Peter A. Browne was one of the earliest hair classifiers in America. A self-proclaimed scientist and a lawyer from Philadelphia, he

[6] Lucile E. Hoyme, "Physical Anthropology and Its Instruments," *Southwestern Journal of Anthropology*, IX (1953), 408–30; Harry L. Shapiro, "The History and Development of Physical Anthropology," *American Anthropologist*, n.s., LXI (June, 1959), 371–79.

[7] Herbert Spencer, *An Autobiography*, 2 vols. (London, 1926), I, 540–42.

[8] Snyder, *Race*, 15; Topinard, *Anthropology*, 350–53; Franz Pruner-Bey, "On Human Hair as a Race Character," Anthropological Institute, London, *Journal*, VI (1876), 71–92.

published his *Classification of Mankind by the Hair and Wool of Their Heads* in 1852, under the "Patronage of the Commonwealth of Pennsylvania." He distinguished three species of men by the horizontal cut of their "pile."[9] The lawyer-scientist lectured in Richmond, Virginia, and Charleston, South Carolina, on the separate species of mankind and won the physician and slavery enthusiast Josiah C. Nott as one of his converts to hair classification. Browne even used his theory in the courthouse of Philadelphia to determine the sanity of an individual according to the dimension of head hairs submitted to his trigameter.[10] Browne's classificatory schema was too obvious in its intent in the 1850's, and though the supporter of slavery, Josiah Nott, listened to his theory, others, critical of his intent, became vocal in their opposition. "By three sweeps of his Discotome," wrote one critic of Browne's schema, he "caught the darkey by the wool." Seizing "time by the forelock," Browne built a theory of hair classification which "led him to seek out a new reason for the National Crime for enslaving the negro-man, in an endeavor to produce evidence that the negro is a separate species."[11] Yet despite the social and political manifestations evident in Browne's classification, his methodology was referred to by later Civil War anthropometric investigations undertaken by the Provost Marshal–General's Bureau.[12] His theory was also used by the medical doctor and anti-abolitionist John H. Van Evrie, who argued not only the Negro's permanent racial inferiority from hair criteria but also that because of the peculiar "matted" nature of the Negro's hair, hats were of purely ornamental value

[9] Peter A. Browne, *The Classification of Mankind by the Hair and Wool of Their Heads* (Philadelphia, 1852), 8–11; Browne, *Trichographia Mammaliam: Or Descriptions and Drawings of the Hairs of the Mammalia, Made with the Aid of a Microscope* (Philadelphia, 1848), 1, 19–20; James Hunt, "On the Negro's Place in Nature," Anthropological Society of London, *Memoirs*, I (1863), 22; "Hair and Wool of the Different Species of Man," *U.S. Magazine and Democratic Review*, XXVII (Nov., 1850), 451–56.

[10] Cornelius G. Peeples, *A New Science to Sustain Slavery* (New York, 1856), 3.

[11] *Ibid.*, 6, 8.

[12] J. H. Baxter, *Statistics, Medical and Anthropological, of the Provost Marshal–General's Bureau, Derived from Records of the Examination for Military Service in the Armies of the United States during the Late War of the Rebellion, of over a Million Recruits, Drafted Men, Substitutes, and Enrolled Men*, 2 vols. (Washington, D.C., 1875), I, 61.

and had no utilitarian purpose, as in the Caucasian's need for protection from the sun.[13]

The facial angle was the most extensively elaborated and artlessly abused criterion for racial somatology. To compare the races, Petrus Camper (1722–1789) had suggested the facial angle. Basically it was a "horizontal line . . . drawn through the lower part of the nose . . . and the orifice of the ear." The angle formed by this horizontal line and the characteristic line of the face made up the facial angle. Using this index, Camper arranged the forms of crania. "The two extremities . . . of the facial line are from 70 to 100 degrees," he wrote, "from the negro to the Crecian antique; make it under 70, and you describe an orang or an ape; lessen it still more, and you have the head of a dog."[14]

The facial angle, used as early as Aristotle as an indication of intelligence, showed a distinct gradation and an implicit manifestation of inferiority. The Greek sculptors, in representing the superhuman attributes of their gods, gave the deities a facial angle of 100 degrees, exceeding that of the highest human. The artistic ideal that gave substance to the statuary scale of beauty in Greek art was incorporated in the speculations of later ethnologists and guided the theory of Camper's facial angle as well as the cranial beauty of Blumenbach's Caucasian skull.[15] When Petrus Camper restored the idea of the facial angle in 1784, "the scientific world gave it a cordial welcome."[16] "The idea of stupidity is associated,

[13] John H. Van Evrie, *White Supremacy and Negro Subordination* (New York, 1868), 100–101.

[14] Petrus Camper quoted in J. S. Slotkin, "Racial Classifications of the 17th and 18th Centuries," Wisconsin Academy of Science, *Transactions*, XXXVI (1944), 465; Camper, *The Works of the Late Professor Camper, on the Connexion between the Science of Anatomy and the Arts of Drawing, Painting, Statuary . . .* (London, 1821), 1–8; Ransom Dexter, "The Facial Angle," *Popular Science Monthly*, IV (Mar., 1874), 587–92; William Lawrence, *Lectures on the Comparative Anatomy and the Natural History of Man* (London, 1840), 246; John C. Greene, *The Death of Adam: Evolution and Its Impact on Western Thought* (Ames, Iowa, 1959), 190–92.

[15] Daniel Wilson, "Ethical Forms and Undesigned Artificial Distortions of the Human Cranium," *Canadian Journal*, n.s., XLI (Sept., 1862), 399–400; Eugene S. Talbot, *Degeneracy, Its Causes, Signs, and Results* (London, 1899), 181–83.

[16] J. W. Redfield, "Measures of Mental Capacity," *Popular Science Monthly*, V (May, 1874), 72; Arthur de Gobineau, *The Inequality of the Human Races* (New York, 1915), 108–9.

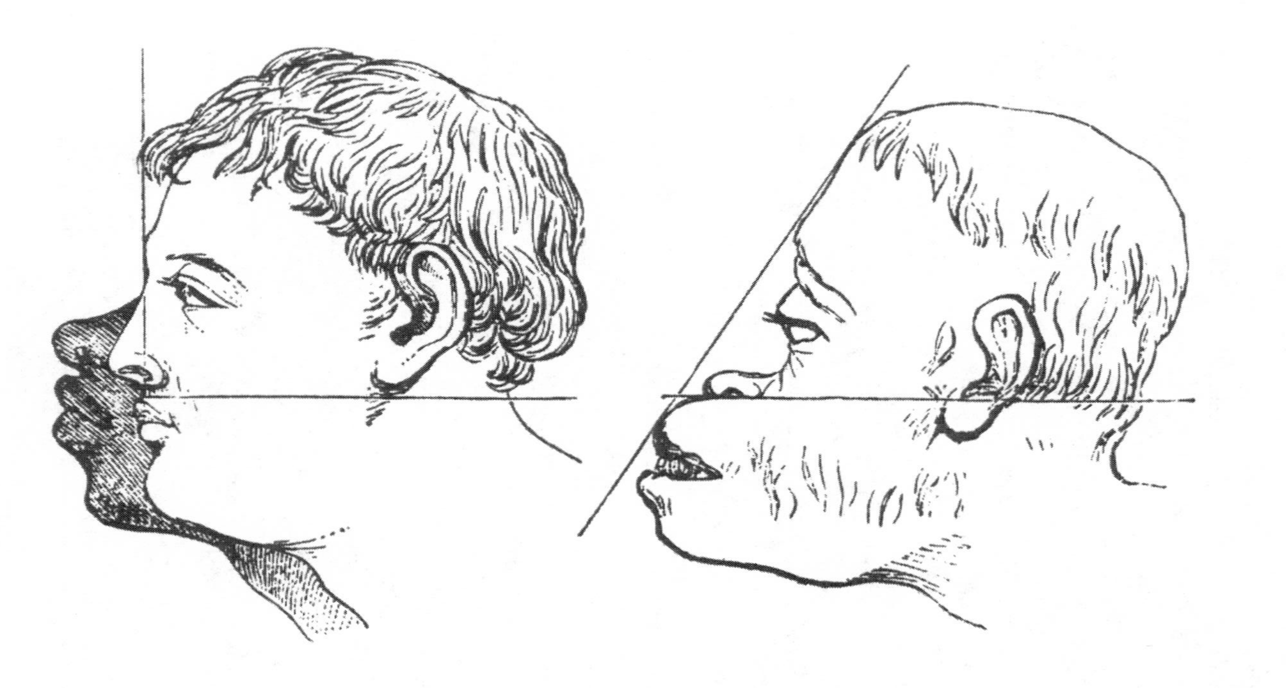

[*Profile of Negro, European, and Oran Outan.*]

Facial angle before publication of Darwin's *Origin of Species* (from Robert Knox, *The Races of Men* [1850]).

Courtesy of National Library of Medicine

even by the vulgar," wrote Camper, "with the elongation of the snout, which necessarily lowers the facial line." For this reason the elephant and the owl were credited with "a particular air of intelligence." The owl was frequently "the emblem of the goddess of wisdom," while the Malayan distinguished the elephant "by a name which indicates an opinion that he participates with men in the most distinguishing characteristic, the possession of reason."[17]

Anders Adolph Retzius in 1840 modified and elaborated upon the early experiments of Camper as well as the experiments of Louis Jean Daubenton (1716–1799) and Lambert Adolphe Jacques Quetelet (1796–1874). Quetelet, whose statistical theories became the basis of the American Civil War anthropometric investigations, believed that the facial angle showed a direct relationship between the proportion of intelligence and the "type or standard of beautiful for the human species," a relationship which gave pre-eminence to the Caucasian.[18] Similar measurements on the facial angle were begun by Walter Barclay and Marcel de Serres, while Friedrich Tiedemann (1781–1861), Samuel Morton, and Mikhail M. Volkoff experimented with internal skull capacity by the use of millet seed, shot, and water weight. By 1860 the facial angle had become the most frequent means of explaining the gradation of species. Like the Chain of Being, the races of man consisted of an ordered hierarchy in which the Hottentot, the Kaffir, the Chinaman, and the Indian held a specific position in the order of life. The only real difference between the pre-Darwinian and post-Darwinian idea of facial angle was the element of evolution. This difference, however, had no bearing on the idea of inferiority inherent in the ordered hierarchy.[19]

Phrenology became another means of racial classification for the

[17] Camper quoted in John Kennedy, *The Natural History of Man* (London, 1851), 17.

[18] Lambert A. J. Quetelet, *A Treatise on Man and the Development of His Faculties* (Edinburgh, 1842), 98.

[19] Winthrop D. Jordan, *White over Black: American Attitudes toward the Negro, 1550–1812* (Chapel Hill, N.C., 1968), chap. 13; Lawrence Johnson, "The Chain of Species," *Popular Science Monthly*, V (July, 1874), 313–22; Andrew Combe, "Remarks on Tiedemann's Comparison of the Negro Brain and Intellect with Those of the European," *Eclectic Journal of Medicine*, II (1837–1838), 325–28.

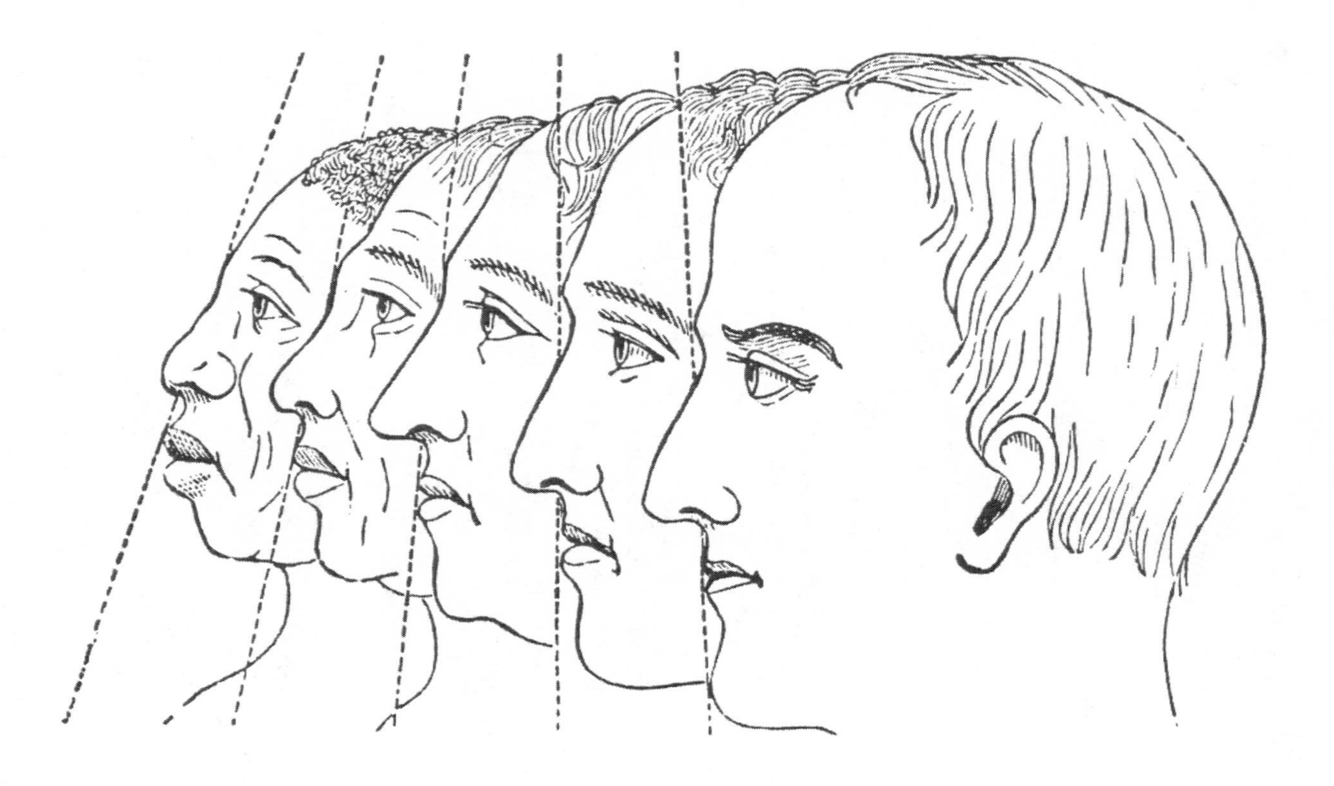

Facial angle after publication of Darwin's *Origin of Species* (from John J. Jeffries, *Natural History of the Human Races* [1869]).

Courtesy of Indiana University

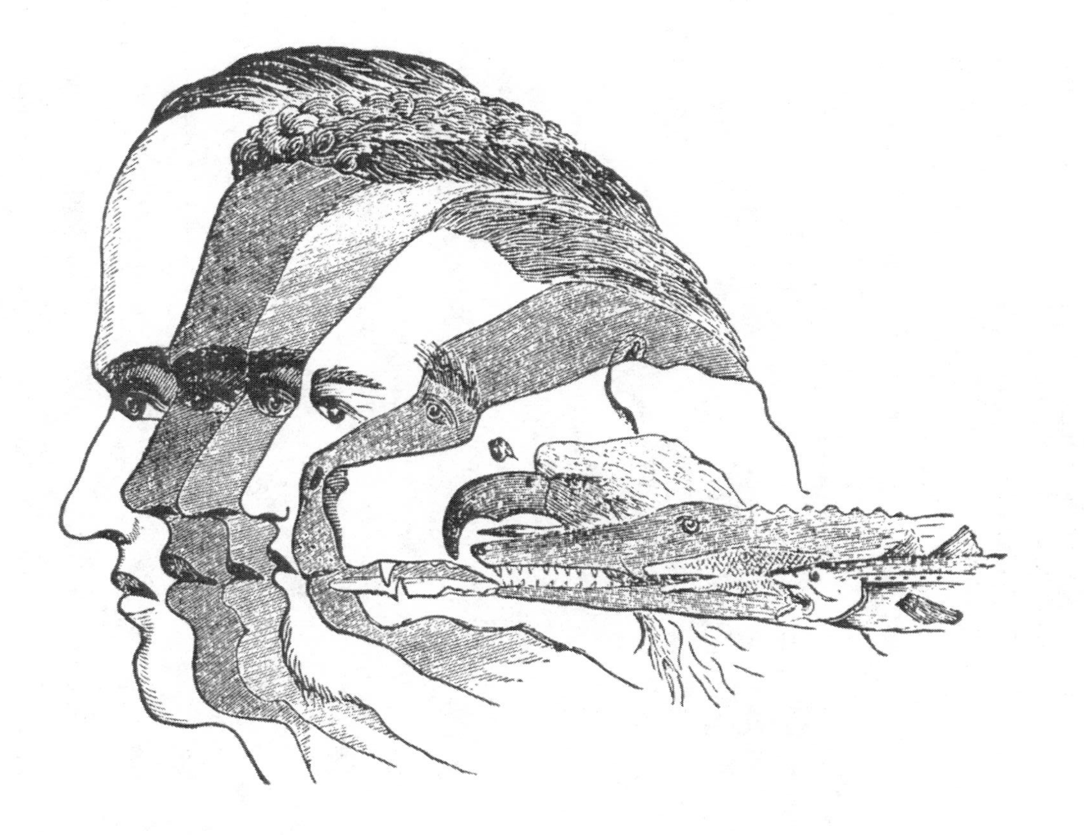

Facial angle after publication of Darwin's *Origin of Species* (from Ranson Dexter, "The Facial Angle," *Popular Science Monthly* [1874]). **Courtesy of Indiana University**

nineteenth-century science of man. It was a science that endeavored to study states of consciousness manifested in "bumps" on the surface of the skull. Later phrenology became a study of the various provinces of the brain, each representing seats of mental aptitudes. Phrenology was incorporated in the Comtean schema of the sciences and, as such, was an effort to bridge the void between the strictly physical aspects of biology and the generalizations concerning man's mental tendencies made from sociology. From the evidence of skull differences, the statistical generalities made from the measurements of particular stocks, nationalities, or races, phrenology offered an easy transition from biology to the suggestive classification of man. It also became a vehicle for the generalities of sociology. Phrenology was for the generation of Comte what Herbert Spencer's psychology became for the era of evolutionism.[20]

After a faddish period of phrenology begun by Franz Joseph Gall about 1800, the real spokesman for it as a scientific part of anthropology (craniometry) was Paul Broca (1824–1880), inventor of the cephalic index, "the breadth of the head above the ears expressed in percentage of its length from forehead to back": "Assuming that the length is 100, the width is expressed as a fraction of it. As the head becomes proportionally broader—that is, the more fully rounded, viewed from the top down—this cephalic index increases. When it rises from 80, the head is called brachycephalic; when it falls below 75, the term dolichocephalic is applied to it. Indexes between 75 and 80 are characterized as mesocephalic."[21]

[20] John Fiske, *The Outlines of Cosmic Philosophy*, vols. I–IV of *John Fiske's Miscellaneous Writings*, 12 vols. (Boston, 1902), IV, 107–9; James Hunt, "On Physio-Anthropology," Anthropological Society of London, *Journal*, V (1867), cclx–cclxxi; J. W. Jackson, "Anthropology and Phrenology," *Anthropological Review*, V (Jan., 1867), 71–79; G. O. Groom Napier, "On the Moral and Intellectual Characteristics of Man," Anthropological Society of London, *Journal*, V (1867), clx–clxix; Robert E. Riegel, "The Introduction of Phrenology to the United States," *American Historical Review*, XXXIX (1933–1934), 73–78; John D. Davies, *Phrenology, Fad and Science* (New Haven, Conn., 1955), 145–48.

[21] William Z. Ripley, "The Racial Geography of Europe," *Popular Science Monthly*, L (Mar., 1897), 577–78; J. G. Garson, "The Cephalic Index," Anthropological Institute, London, *Journal*, XVI (1886–1887), 11–17; John Knott, "Franz Josef Gall and the Science of Phrenology," *Westminster Review*, CLXVI (1906), 150–63.

Born in 1824 of Huguenot background, Broca became active in the Society of Surgery and the Anatomical Society of Paris, and was later one of the founders of the Society of Anthropology of Paris in 1859. Broca invented many of the instruments of craniometry, including the craniograph in 1860, the new goniometer which measured the facial angle in 1864, the stereograph in 1867, the *cadre à maxima* and micrometric compass in 1869, and the occipital goniometer, which measured the angle of the back of the skull, in 1870. He admitted to the difficulties of the science by citing English naturalist Philip Henry Gosse (1810–1888), who in a paper of 1855 discussed the artificial deformation of the skulls of children among the Chinook Indians, the Quechuas and Aymara natives of Peru, and the Koskeemo of Vancouver Island as producing fictitious shapes that would confound craniological determinations. Broca also admitted that diseases in childhood could cause deformation of the skull. Yet he remained adamant about the usefulness of craniometry. He asserted that it was possible for one "to detect the primitive type in a deformed cranium" since quite often the deformed cranium reflected, in an exaggerated form, those characteristic features which the savage admired in his ancestors.[22] Like a collector of the eighteenth century, Broca sought solutions to the past ages in the nearly 500 crania in the society's museum, and at his death he had accumulated over 180,000 measurements.[23]

Despite the assertions of Broca, many somatometrists were skeptical of craniometry. Significantly, some of these skeptics had other than purely scientific observation in their intent. When, for example, they discovered that the shape of the skulls of Negroes and

[22] Paul Fletcher, "Paul Broca and the French School of Anthropology," in Anthropological and Biological Societies of Washington, *Saturday Lectures* (Washington, D.C., 1882), 130–31; Gudmund Hatt, "Artificial Moulding of Infant's Head among the Scandinavian Lapps," *American Anthropologist*, n.s., XVII (Apr.-June, 1915), 245–56; Henry R. Schoolcraft, *Historical and Statistical Information Respecting the History, Condition and Prospects of the Indian Tribes of the United States*, 6 vols. (Philadelphia, 1851–1857), II, 325; William H. Flower, "Fashion in Deformity," *Popular Science Monthly*, XVII (Oct., 1880), 721–42; "Review of the Proceedings of the Anthropological Society of Paris," *Anthropological Review*, I (Aug., 1863), 287; Wilson, "Ethical Forms," 417, 419, 423, 426.

[23] "Review of the Proceedings of the Anthropological Society of Paris," 291; Fletcher, "Paul Broca," 139.

Scandinavians were similar enough to classify them in the same race grouping, it was apparent to many that something was wrong.[24] However, a large part of the skepticism grew not so much from the somatometry of the science as it did from those enthusiasts who sought to go behind the measurements to judge moral character, intelligence, and social tendencies. The measuring devices, developed as a result of the science, outlasted the premises that called them into existence. While phrenology bowed to the harsh criticism of the late nineteenth century, craniometry and its measuring devices survived. Harvard professor Roland Dixon (1875–1934) and Eugene Pittard (1867–1938) were still challenging the use of phrenology as late as the 1920's, and while William Z. Ripley (1867–1941) of America used the cephalic index in his *Races of Europe* (1899), he still felt it necessary to caution strongly against drawing moral conclusions from the measurements.[25] The difficulty in craniometry was not so much the anthropometry involved as it was the premises upon which the science was built and which were used to explain size and shape differences. As a factor in helping to identify unmixed racial types, the cephalic index was valuable, but efforts to connect head form with intellectual power proved damaging to the entire framework of craniometry.[26]

Though phrenologists and craniometrists gradually admitted the limitations of head size and brain weight in determining intelligence, they refused to discount the statistics altogether from their evidence. In fact, this situation seemed to pervade the whole of anthropometry in the nineteenth century. Attacked or criticized on

[24] R. S. Woodworth, "Racial Differences in Mental Traits," *Science*, XXXI (Feb., 1910), 171–86; "Anthropology," *American Naturalist*, VII (Feb., 1870), 117–18; Fiske, *Outlines of Cosmic Philosophy*, III, 195–97; Robert D. Simons, *The Colour of the Skin in Human Relations* (New York, 1961), 1.

[25] William Z. Ripley, *The Races of Europe: A Sociological Study* (New York, 1937), 39; Roland B. Dixon, *The Racial History of Man* (London, 1923), 8; Snyder, *Race*, 11; Allen Starr, "The Old and New Phrenology," *Popular Science Monthly*, XXXV (Oct., 1889), 730; S. Washburn, "Thinking about Race," Smithsonian Institution, *Annual Report for 1945*, 375.

[26] Joseph Simms, "Brain Weight and Intellectual Capacity," *Popular Science Monthly*, L (Dec., 1898), 243–55.

a specific measurement as a criterion for racial comparison, the anthropometrist usually compromised the point at issue and turned to yet another measurement. It was as if the whole was greater than the sum of the parts, as if the accumulation of inferences or measurements, however questionable in themselves, would produce a scientific truth.[27]

By 1860 many of the century's naturalists were leaving phrenology to cranks and outdated enthusiasts of Comte, while placing more and more credence in the new evolutionary psychology as a valid means of determining intelligence. Yet the transition from phrenology to the psychology of Herbert Spencer was neither distinct nor, for that matter, ever really clarified in the community of anthropologists. This situation was exemplified in the continuity of race concepts developed during the heyday of phrenology, which were assimilated without notice into the vocabulary of the evolutionists. Phrenology died a pauper's death in the late nineteenth century, victimized by the vicious ostracism of the period's most reputable anthropologists. But race classification, begun or "proven" by the phrenologists, relegating Mongolian, Malayan, Indian, and Ethiopian to inferior roles beneath the Caucasian, was seldom criticized.[28]

The claims of men like Franz Gall (1758–1828), Charles Bray (1811–1884), John Jackson (1835–1911), and Paul Broca stimulated a prodigious interest in the comparative measurements of the size of the brain. As the science of man grew more professionalized, would-be phrenologists moved from exterior skull measurements to skull capacity. The brain cavity was the focus of many anthropological studies of Samuel Morton, Paul Broca, Joseph Virey, Friedrich Tiedemann, Theodor Welcker, and J. Barnard Davis. The materials used in gauging cranial capacity varied from mercury to sand, white mustard seed, pearl barley, shot, water,

[27] D. Kerfoot Shute, "Racial Anatomical Peculiarities," *American Anthropologist*, o.s., IX (Apr., 1896), 123–32; Topinard, *Anthropology*, 229; C. M. Poynter, "Some Conclusions Based on Studies in Cerebral Anthropology," *American Anthropologist*, n.s., XIX (Oct.-Dec., 1917), 496–97.

[28] Eric T. Carlson, "The Influence of Phrenology on Early American Psychiatry," *American Journal of Psychiatry*, CXV (Dec., 1958), 535–38.

and rubber bags.[29] Jeffries Wyman of the Boston Society of Natural History demonstrated the difficulties involved in measuring the capacity of the cranial cavity when he measured a single skull with eight different materials and obtained eight different measurements varying from 1,193.0 to 1,313.0 cubic centimeters.[30] Yet, despite the difficulties involved in cranial measurements, anthropologists assumed the existence of a uniform relationship between the size of the skull and the development of the intellectual faculties, a relationship which resulted in a graduated series of skull measurements from the anthropoid through the various stages in savage man, culminating in the most civilized nations. Beginning with Australians, Hottentots, and Polynesians and moving slowly up the ladder into the civilized nations, cranial capacity corresponded directly with the degree of civilization achieved. Taking the European skull as the basis of index, the races of man presented an ascending scale of cubic capacity.[31]

RACE	AUTHORITY	CAPACITY
European	Tiedemann	100.0
Asiatic	Davis	94.3
African	Davis	93.0
American	Tiedemann	95.0
American	Davis	94.7
American	Morton	87.0
Oceanic	Davis	96.9

[29] Topinard, *Anthropology*, 116–17, 230—31; Washington Matthews, "Use of Rubber Bags in Gauging Cranial Capacity," *American Anthropologist*, o.s., XI (June, 1898), 171–76; G. Busk, "Ready Method of Measuring the Cubic Capacity of Skulls," Anthropological Institute, London, *Journal*, III (1873), 200–204; Owsei Temkin, "Gall and the Phrenological Movement," *Bulletin of the History of Medicine*, XXI (1947), 275–331; "The Weight and Development of the Brain as Indicative of Intellectual Force," *American Journal of Insanity*, LIX (Apr., 1883), 471–75; Curtin, *The Image of Africa*, 366–69; Samuel Morton, "Observations on the Size of the Brain in Various Races and Families of Man," Philadelphia Academy of Natural Sciences, *Proceedings*, IV (1848–1849), 221–24; J. Barnard Davis, "Contributions toward Determining the Weight of the Brain in the Different Races of Man," Royal Society of London, *Proceedings*, XVI (1867–1868), 236–41; F. Peterson, "Some of the Principles of Craniometry," *Medical Record* (New York), XXXIII (1888), 681–86.

[30] Daniel Wilson, "Brain-Weight and Size in Relation to Relative Capacity of the Races," *Canadian Journal*, n.s., XCII (Oct., 1876), 182.

[31] *Ibid.*, 201.

RACE	AUTHORITY	CAPACITY
Chinese	Davis	99.8
Mongol	Morton	94.0
Mongol	Tiedemann	93.0
"Hindoo"	Davis	89.4
Malay	Tiedemann	89.0
American Indian	Morton	91.0
"Esquimaux"	Davis	98.8
Mexican	Morton	88.5
Peruvian	Wyman	81.2
Peruvian	Morton	81.2
Negro	Tiedemann	91.0
Negro	Peacock	88.0
Hottentot	Morton	86.0
Javan	Davis	94.8
Tasmanian	Davis	88.0
Australian	Morton	88.0
Australian	Davis	87.9

CIVIL WAR ANTHROPOMETRY: THE MAKING OF A RACIAL IDEOLOGY

The Civil War in America stands as a watershed in nineteenth-century anthropometric developments. Body measurements collected during the war years marked the culmination of efforts to measure the various "races" or "species" of man and derive a semblance of understanding as to specific racial types. Both the Provost Marshal–General's Bureau and the United States Sanitary Commission, a semiofficial organization made up of "predominantly upper-class . . . patrician elements which had been vainly seeking a function in American society" during the Civil War, became pioneer forces in the wide-scale measurement of the soldier during the war years.[32] The war marks a watershed not so much because its conclusions were new but because nearly all subsequent late nineteenth-century institutionalized attitudes of

[32] George M. Fredrickson, *The Inner Civil War: Northern Intellectuals and the Crisis of the Union* (New York, 1965), 100.

racial inferiority focused upon war anthropometry as the basis for their beliefs. Ironically, the war which freed the slave also helped to justify racial attitudes of nineteenth-century society. The direction and conclusions of the Civil War anthropometric evidence buttressed the conservative ethos of American social order and stability and, at the same time, encouraged a new "scientific" attitude.

The reason the Civil War became such an important catalyst in the development of anthropometry stemmed from two particularly troublesome wartime situations. First, as a result of the embarrassing Union defeat in the first battle of Bull Run, Lincoln authorized on June 13, 1861, the creation of the United States Sanitary Commission. Its function was to make a study of the physical and moral condition of federal troops, carry out anthropometric examinations of soldiers, and offer suggestions and aid for improvements in army life. The life insurance companies of America underwrote a large portion of the commission's expenses, since they were willing to subsidize almost any program that could work out statistical averages on the physical condition of the population.[33] Members of the commission included Henry W. Bellows, Unitarian minister of New York, Alexander Dallas Bache of the Coast Survey, Dr. Wolcott Gibbs of Massachusetts, Dr. Samuel Gridley Howe, educator and philanthropist, Dr. William H. Van Buren of New York, and Charles J. Stillé, lawyer and historian of the Sanitary Commission. Frederick Law Olmstead became the general secretary of the Sanitary Commission and, while it operated independently of the federal army, it was subject to the prerogatives of the Secretary of War, Edwin Stanton. A second situation, and one which became extremely important to the anthropometric section of the Sanitary Commission, grew out of the July 17, 1862, congressional authorization for Lincoln "to employ as many persons of African descent as he may deem necessary and proper for the suppression of the Rebellion." The act permitted Lincoln to use Negroes in "any military or naval service that they

[33] Charles J. Stillé, *History of the United States Sanitary Commission* (Philadelphia, 1866), 84.

may be found competent." Eventually over 180,000 Negroes were inducted into the federal army.[34]

European anthropologists had made studies on groups of individuals before the American Civil War, but their findings were not very comprehensive. John Towne Danson (1817–1898) took measurements of some 733 Liverpool prisoners of all ages, James David Forbes (1809–1868) of Scottish students at Edinburgh, and Franz Liharzik (1813–1866) of 300 Viennese men.[35] There were also extensive measurements made during the Crimean War. But European anthropological societies as well as interested numbers of American scientists looked upon the creation of the Sanitary Commission, and the induction of Negroes into the Union Army, as an opportune means of investigating race differences on a scale never before achieved. Somatological differences, which previously had been ascertained from random measurements upon small numbers and with a variety of measuring devices, could now be taken on a wide scale, with planned experiments and uniform measuring instruments.

The Sanitary Commission based its anthropometric investigations upon the statistical methodology of the Belgian philosopher Lambert Quetelet. Quetelet had made several statistical analyses of human physiognomy, including examinations of 900 men enrolled for draft in Brussels, 9,500 Belgian militia, 69 convicts in a penitentiary at Vilvarde, and 80 students at Cambridge, England. In 1846 Quetelet applied his theory of probability to "moral and political science," and his results were given wide audience by Sir John Herschel. Herschel's extended article in the *Edinburgh Review* on Quetelet's methodology "led the way to examination of the subject in Great Britain, and, later, in the United States."[36]

[34] Quoted in Sanford B. Hunt, "The Negro as a Soldier," *Anthropological Review*, VII (Jan., 1869), 41; "The Sanitary Commission," *North American Review*, XCVII (Jan., 1864), 167, and (Apr., 1864), 370–419; Benjamin A. Gould, *Investigations in the Military and Anthropological Statistics of American Soldiers* (New York, 1869), 14.

[35] Gould, *Investigations*, 119; Royal Statistical Society of London, *Journal*, XXV (1862), 24.

[36] Baxter, *Statistics, Medical and Anthropological*, I, lxxvii; Ezekiel B. Elliott, *On the Military Statistics of the United States of America* (Berlin, 1863), 14–15.

The whole basis of Quetelet's researches was the creation of an "average man" as representative of specific groups and an analysis of that specimen "in his various relations, physical, social and moral."[37]

Drawing statistics and relationships out of a multitude of examinations of soldiers, the Sanitary Commission sought to construct Quetelet's average man. In finding him among the "native American," British American, English, Irish, German, "foreigner," Negro, Indian, and "college student," the commission determined profiles of an abstract man to whom they assigned a statistical intellect, capacity, judgment, and tendency. It was a study oriented from its very inception upon a proper understanding of the varieties of man—a reflection of the reformer's zeal in the early years of anthropology in America.

> Indeed the external form of this average man may legitimately be adopted as a standard of beauty and a model for art. The eminent scientist already named [Quetelet] has shown that we may discover not merely the outward semblance of this abstract being, but his needs, capacities, intellect, judgement, and tendencies; and Quetelet may thus be regarded as the founder of statistical anthropology, indeed of social science, in the true significance of the word, according to which science depends upon the investigation of laws, not upon the consideration of isolated facts, nor the dissemination of correct principles.[38]

In July, 1864, the Sanitary Commission invited Benjamin A. Gould, a member of the National Academy of Sciences and president of the American Association for the Advancement of Science, to assume direction of extension of the anthropometric statistics undertaken in 1863 by Ezekiel B. Elliott, the commission's first actuary.[39] In the reports of the Sanitary Commission published in

[37] Gould, *Investigations*, 244; Howard Becker and Harry E. Barnes, *Social Thought from Lore to Science*, 2 vols. (Washington, D.C., 1952), I, 563; Quetelet, *A Treatise on Man*, 74; Franz Boas, "Remarks on the Theory of Anthropometry," American Statistical Association, *Proceedings*, III (Dec., 1893), 569–75.

[38] Gould, *Investigations*, 246; Edward B. Tylor, "Quetelet on the Science of Man," *Popular Science Monthly*, I (May, 1872), 45–55.

[39] Gould, *Investigations*, v; Erving Winslow, "Sketch of Professor Benjamin Gould," *Popular Science Monthly*, II (Mar., 1882), 683–87; Ezekiel B. Elliott, *Preliminary Report on the Mortality and Sickness of Volunteer Forces of the United States Government during the Present War* (New York, 1862); "Death of E. B. Elliott," *Science*, XI (June, 1888), 261.

1869, Gould admitted freely to a variety of difficulties encountered in the investigations. For one thing, Secretary of War Stanton had continually declined to assist the commission in its efforts to obtain information. He denied them use of War Department records and hindered plans for more extensive investigations. Part of the explanation for Stanton's attitude was that a similar military anthropometric study had been inaugurated by the Provost Marshal-General's Bureau in 1861. Perhaps Stanton declined to aid the commission because of departmental pressure from the Provost Marshal.[40] In any case Gould took every opportunity in the published reports to remark on Stanton's unwillingness to help them. Other difficulties that the anthropometric section members experienced grew from the lack of intelligent classification. This situation became evident in their unsuccessful attempts to define adequately various mixtures of Negro blood, in the realization that they had made statistical studies of an Iroquois tribe only and yet were speaking of the Indian in general, that they were unaware of the number of mixed-blood Iroquois they had examined, and that quite often accidental errors occurred in examination procedures. Procedural errors were most evident in the confusion surrounding use of the facial angle instrument. Unfortunately, much of the statistical data taken during the Civil War "was carried out under unfavorable circumstances and by men many of whom had no previous knowledge of these matters, and who received no instruction except by circulars."[41]

The instruments used by the commission—andrometer, spirometer, dynomometer, facial angle, platform balance, calipers, and measuring tape—were intended to measure "the most important physical dimensions and personal characteristics." Fortunately for the commission, Joseph Henry (1797–1878), first secretary of the Smithsonian Institution, had undertaken similar studies a few years earlier. The Sanitary Commission utilized the design of Henry's apparatus in order to facilitate uniformity in instruments

[40] Gould, *Investigations*, 298.

[41] *Ibid.*, 146, 384–97; Aleš Hrdlička, "Physical Anthropology: Its Scope and Aims; Its History and Present Status in America," *American Journal of Physical Anthropology*, I (Apr.-June, 1918), 172.

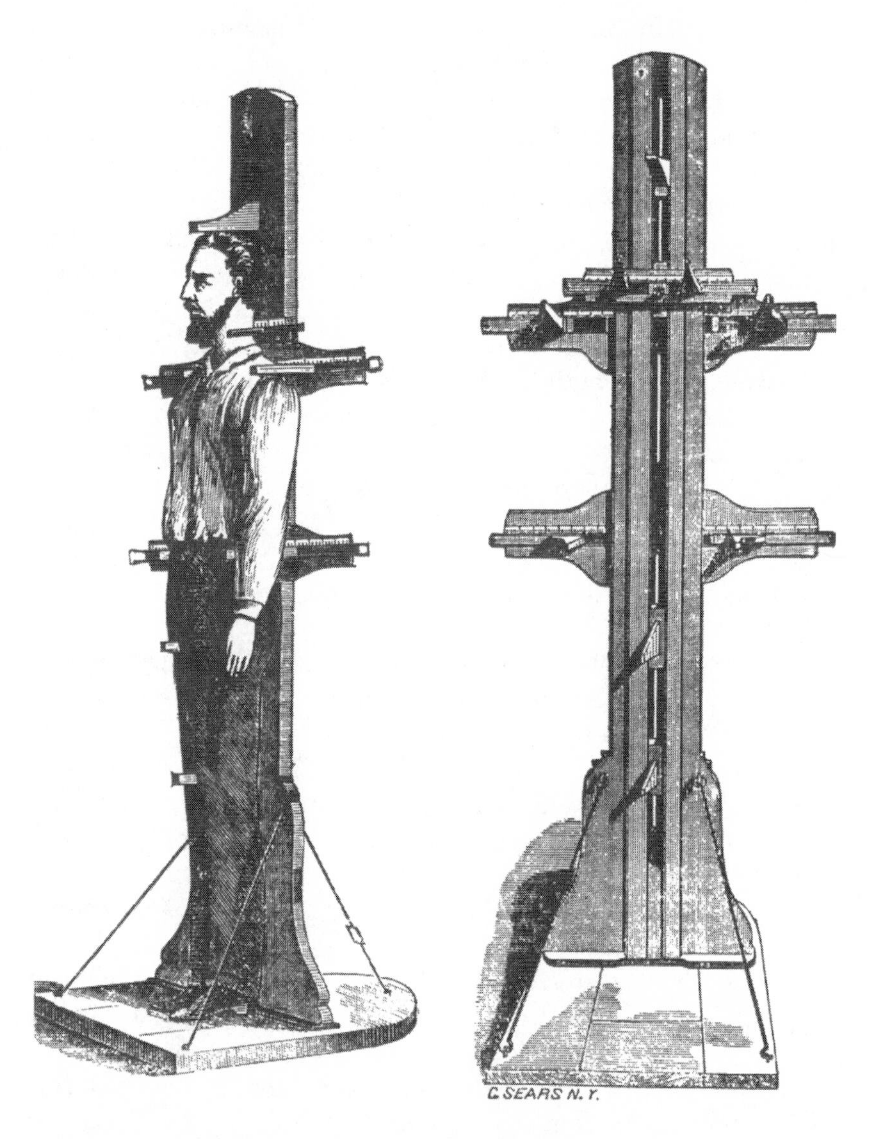

Andrometer used by U.S. Sanitary Commission to determine principal physical measurements (from Benjamin A. Gould, *Investigations in the Military and Anthropological Statistics of American Soldiers* [1869]).

Courtesy of National Library of Medicine

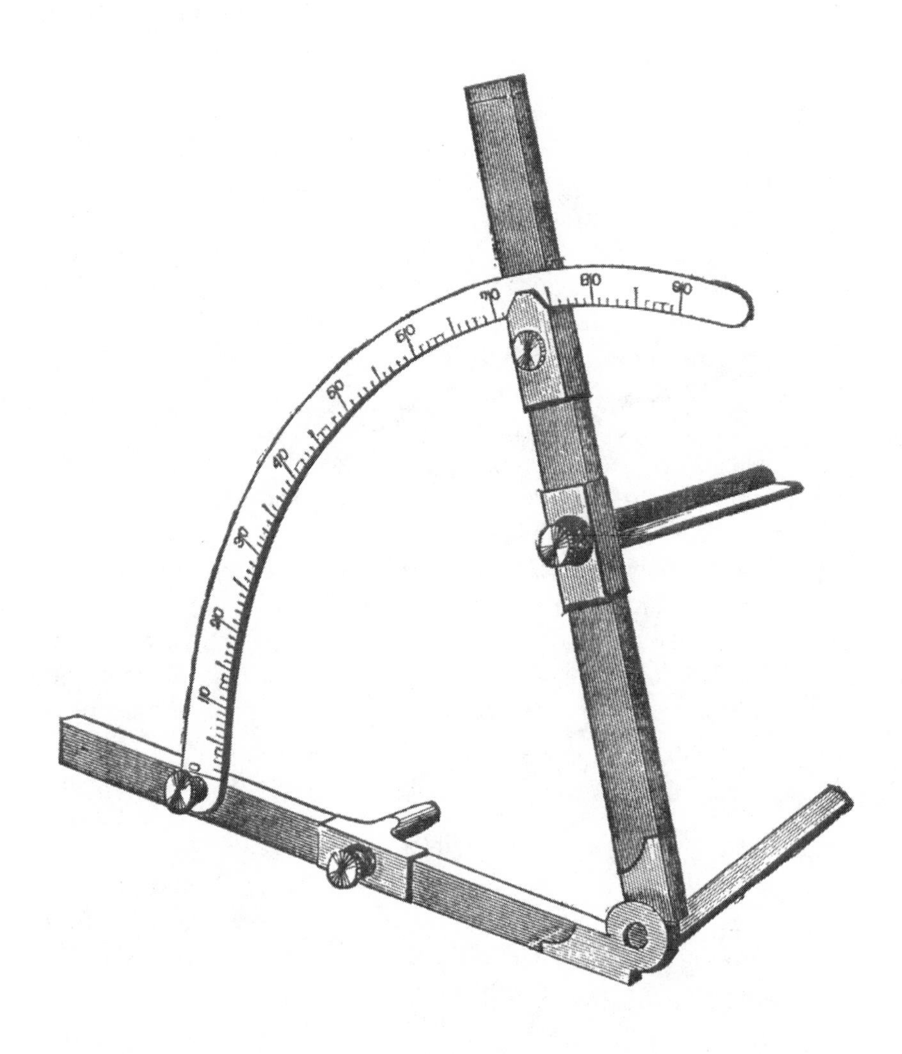

Facial angle instrument used by U.S. Sanitary Commission; it closely resembles the instrument designed by Paul Broca (from Gould, *Investigations*).

Courtesy of National Library of Medicine

and procedure. The Coast Survey office built the remaining instruments under the guidance of Professor Alexander Dallas Bache, vice-president of the commission and superintendent of the Coast Survey.[42]

In the first year of its operations the anthropometric section examined 8,004 white Union troops and rebels. Out of that number the commission accepted 7,904 as valid examinations for statistical analysis. In July, 1864, the commission suggested modifications in both the apparatus and the form (Form E) containing the statistical data requested in the examination. The newly modified Form EE clearly reflected the commission's recognition of Lincoln's call for Negro soldiers. "No examination of the negro troops seem to have been made yet," the commission argued, "and the importance of such inspections needs no comment." Modification of the form meant that information concerning the Negro in America could be ascertained "with advantage."[43] As a result of the suggestions of the commission, the statistical section added six more measurements to the original form: (1) distance from tip of middle finger to level of upper margin of patella, (2) height to knee, (3) girth of neck, (4) perineum to most prominent part of the pubes, (5) distance between nipples, and (6) circumference around hips. The commission also made corrective modifications in the apparatus. With the aid of Louis Agassiz, Jeffries Wyman, William H. Holmes, and J. H. Douglas, the instruments were further refined for closer and more exact measurements.[44]

During the second phase of examination, which lasted until the end of the war, a staff of twelve examiners drew statistics from 15,900 examinations, of which 15,781 were accepted as valid. The total consisted of 10,876 white soldiers, 1,146 white sailors, 68 white marines, 2,020 full-blooded Negroes, 863 mulattoes, and 519 Indians. The examination of Indians, mostly Iroquois, was made while they were held for a time as prisoners of war near Rock Island, Illinois. There were, in addition to these measurements,

[42] Gould, *Investigations*, 218; "Alexander D. Bache," *Appleton's Cyclopaedia for 1867* (New York, 1872), 78–79; Elliott, *On the Military Statistics of the United States*, 10.

[43] Gould, *Investigations*, 221.

[44] *Ibid.*, 218–27.

TABLE IX.

Comparison of Mean Dimensions.

	White Soldiers		Sailors	Students	Full Blacks	Mixed Races	Indians
	Later Series	Earlier Series					
Number of Men . .	10 876	7 904	1 061	291	2 020	863	517
Mean Age	y 26.2	y 25.1	y 26.1	y 21.7	y 25.7	y 26.2	y 30.7
Length Head & Neck	in. 9.944	in. 9.981	in. 10.091	in. 10.098	in. 9.623	in. 9.561	in. 9.547
Length of Body . .	26.140	26.099	24.549	26.109	24.487	24.680	26.870
Knee to Perinæum .	12.456	–	12.880	12.652	12.964	12.692	12.799
Height to Knee . .	18.609	–	18.498	19.240	19.136	19.318	19.009
Stature	67.149	67.366	66.018	68.099	66.210	66.251	68.225
Acromion to Elbow .	13.605	–	13.171	13.712	13.302	13.856	13.757
Elbow to Finger-tip .	15.548	–	15.367	15.309	16.103	16.415	17.035
Dist. betw. Acromia	12.731	16.359[a]	12.879	13.085	14.089	14.742	12.830
Ratio of parts of Arm	1.143	–	1.167	1.116	1.211	1.185	1.238
" " Leg	1.494	–	1.436	1.521	1.476	1.522	1.485
Med. line to Finger-tip	35.042	–	33.848	34.920	35.808	35.822	37.198
Acromion " "	29.153	29.200[b]	28.538	29.021	29.405	30.271	30.792
Height to Perinæum	31.065	31.286	31.378	31.892	32.100	32.010	31.808
Ratio of Leg to Arm	1.066	1.071	1.100	1.099	1.092	1.058	1.033
Height to Pubes . .	–	–	33.269	–	34.302	34.534	–
Finger-tip to Patella	5.036	–	5.778	6.473	2.884	4.125	3.653
Circumf. of Waist .	31.467	32.089	30.457	31.240	30.296	30.546	34.593
Circumf. of Hips .	36.930	–	34.942	36.549	35.569	35.357	38.962
Circumf. of Chest .	35.818	35.353[c]	35.124	35.313	35.087	34.966	38.001
Play of Chest . . .	2.65	–	2.08	3.07	1.62	1.57	1.84
Dist. between Nipples	8.136	–	8.304	8.071	7.970	7.891	–
Ratio to circum. Chest	0.226	–	0.236	0.229	0.225	0.227	–
Dist. between Eyes .	2.492	2.606	2.473	2.484	2.714	2.670	2.716
Breadth of Pelvis .	11.916	13.153[d]	11.625	11.187	10.952	11.267	12.889
Length of Foot . .	10.058	–	10.114	9.957	10.600	10.439	10.123
Thickness of Foot .	2.572	–	2.921	2.786	2.672	2.770	2.687
Length of Heel[e] . .	0.48	–	0.49	0.46	0.82	0.57	0.48

[a] Full breadth of shoulders. [b] Measured from arm-pit.
[c] Not the half-sum of circumferences at inspiration and expiration, as the others are.
[d] Probably the breadth of hips. See page 262.
[e] These values are obtained by adding 0.3 to the difference between the dimensious 36*a* and 36*b*. See page 274.

Anthropometric statistics derived by U.S. Sanitary Commission (from Gould, *Investigations*). *Courtesy of National Library of Medicine*

statistics taken from the examination of three dwarfs and two captured "Australian children." These latter were made almost as an afterthought and had no bearing on the main body of material, though surely they reflected the avid curiosity of nineteenth-century anthropologists about specimens of atavism and savagery.[45]

Bridging a variety of topics, the results of the Sanitary Commission measurements far surpassed any collection previously made. The records of the commission report compared and contrasted various nationalities, college students, Indians, and Negroes according to body dimensions, head size, strength, teeth, vision, respiration, and pulmonary capacity. The evidence offered an immediate refuge for both hereditarians and environmentalists among the anthropologists. For one thing, the report showed that there were perceptible differences between free and slave state Negroes with respect to head size, height, and weight.[46] In its report on the mulatto the statistics were interpreted as corroboration of earlier racialist assertions that the product of miscegenation was physiologically inferior to the original stocks and, therefore, that mixing races was no real remedy to the racial inferiority of the Negro. "The curious and important fact that the mulattoes, or men of mixed race occupy so frequently in the scale of progression a place outside of, rather than intermediate between, those races from the combination of which they have sprung," stated the report, "cannot fail to attract attention. The well-known phenomenon of their inferior vitality may stand, possibly, in some connection with the fact thus brought to light."[47]

Those characteristics which most marked the races, according to the report, were, for the white, "the length of the head and neck and the short fore-arms"; for the Indian, "the long fore-arms and the large lateral dimensions, excepting at the shoulders"; and for the black, "the wide shoulders, long feet, and protruding heels."[48] The length of the forearm was important to the anthropometrist. It was the measurement the commission added to the original

[45] *Ibid.*, 312–15.
[46] *Ibid.*, 147, 297, 347, 379, 568.
[47] *Ibid.*, 319.
[48] *Ibid.*

statistical form for the benefit of Negro and Indian examinations. The measurement applied to the difference found in the distance from the fingertip to the kneepan. Here the full Negro was but three-fifths and the mulatto five-sixths the average distance for the white soldier. This difference was due to the greater arm length and shorter body length of the full black, and marked the Negro as that much closer to the anthropoid in development.[49] The report went on to compare chest size and concluded from its statistics that "the difference between the mulattoes and the full blacks is here very conspicuous . . . the blacks in their turn falling below the Indians, and these vastly below the whites, of whatever class."[50]

After the war the Sanitary Commission, in an effort to continue uniformity in anthropometrics, distributed its apparatus among colleges and institutions for continued research. Also distributed were the modified forms and instructions. Although the commission admitted to defects in the apparatus, forms, and procedures of examination, it felt that the program was a step forward, rendering American anthropometric investigations more uniform than any yet performed and providing a useful, singular guide for future race study.[51]

In 1875 J. H. Baxter brought out *Statistics, Medical and Anthropological, of the Provost Marshal–General's Bureau.* Though varying at times from the conclusions of Gould's Sanitary Commission reports, Baxter's investigations, carried on between 1861 and 1865, generally corroborated on a much larger scale the earlier findings. One of the interesting elements of the army study was a questionnaire sent to military medical doctors requesting their observations of Negro recruits—their physical build, intelligence, and ability to render military service. A large number of doctors refrained from answering the portion relating to the Negro since many of them had few or no Negro recruits upon which to base judgment. Those who did offer remarks gave surprisingly similar conclusions. The Negro in America, by reason of his contact with a higher civi-

[49] *Ibid.*, 347.

[50] *Ibid.*, 359; "The Negroes and Indians of the United States," *Anthropological Review*, IV (Jan., 1866), 40–42.

[51] Gould, *Investigations*, 231; Joseph Henry, "Report of the Secretary," Smithsonian Institution, *Annual Report for 1865*, 47–48.

lization, lost most of his "grosser peculiarities." This factor, along with his good physical endowment, made him a capable soldier.[52] His only apparent physical deformation was his flat feet.[53] Though a good soldier and perhaps a good citizen, wrote Dr. E. S. Barrows of Iowa, the Negro "never can be as well qualified as he who by nature possesses greater physical perfection and greater mental endowments."[54] The smaller facial angle of the Negro recruit, wrote a New Jersey doctor, denoted a physical organization of "brute force rather than intellectual pre-eminence," a situation which relegated him to the lower tasks of society.[55]

Like the conclusions of Gould's Sanitary Commission reports, Baxter's questionnaire to Union doctors confirmed the prevailing belief in the physical inferiority of the mulatto. Negroes of mixed blood were incapable of enduring hardship and were weaker than either the pure black or the white. As a class, wrote Dr. J. H. Mears of Pennsylvania, the colored race "furnished a larger proportion of men who have passed the examination than any other." On the other hand, those rejected were invariably mulattoes.[56] Though imitative, the powers of the mulatto were a good deal less than the full black, and the mulatto exhibited a greater tendency to scrofulous disorders.[57]

In 1869 Sanford B. Hunt, who was a surgeon in the United States Volunteers, published an article in the London *Anthropological Review* entitled "The Negro as a Soldier." The article was a copy of a report he had made to the United States Sanitary Commission and had been published with the permission of Dr. William A. Hammond, Surgeon-General of the United States Army. The conclusions which Hunt felt could now be ascertained about the Negro concerned such things as his "capacity to learn tactics," personal hygiene, "powers of resistance to hunger and fatigue," diseases, morale, courage, obedience, cheerfulness, and "his comparative intellectuality." The "well known imitative faculty" of the

[52] Baxter, *Statistics, Medical and Anthropological*, I, 334, 370, 394, 465.
[53] *Ibid.*, 311, 394.
[54] *Ibid.*, 461.
[55] *Ibid.*, 285.
[56] *Ibid.*, 394, 403.
[57] *Ibid.*, 285.

Negro, along with "his natural fondness for rhythmical movement," made him a good recruit for the drill-master. "The habit of obedience inculcated by the daily life of the slave," added to the Negro's ability to become a worthy soldier. His "large, flat, inelastic foot . . . almost splay-footed," gave him an advantage in marching over rough terrain.[58]

"It would be grossly unfair to subject the negro," argued Hunt, "to a comparison of intellectual capacity based on his present manifestations of mental acuteness." Held in ignorance by the southern planter, he was barred from education and all paths of competition. Hence his inferiority, being of an environmental sort, blunted any mental test that might be used to define his relative position in the scale of races. For this reason Hunt suggested three different modes of determination: (1) by "external measurements of the cranium," (2) by ascertaining a direct ratio "between the mental and the cubic capacity of the cerebral mass," attempted before the war by Samuel George Morton, and (3) by determining the "weight of the brain by post-mortem examinations." Of the three possible methods, Hunt chose the last as being more reliable. All three methods, he admitted, "presuppose that the size and weight of the brain is the measure of its intellectuality."[59]

Hunt had made studies of the autopsies performed during the Civil War at Benton Barracks in Missouri, Wilson Hospital in Nashville, Tennessee, and L'Ouverture Hospital in Alexandria, Virginia. He drew up statistics derived from 405 autopsies of white and Negro soldiers made under the direction of surgeon Ira Russell of the 11th Massachusetts Volunteers. Twenty-four of the autopsies were performed on white soldiers and 381 on blacks. Hunt concluded from brain-weight analysis that the full-blooded Negro brain weighed five ounces less than the white, that "slight intermixtures" of white blood in the Negro "diminish the negro brain from its normal standard" while large infusions of white blood, such as in the mulatto, "determine a positive increase in the negro brain, which in the quadroon is only three ounces below the white standard." Though the statistics of the Sanitary Commission autop-

[58] Hunt, "The Negro as a Soldier," 42–43.
[59] *Ibid.*, 49–50.

sies showed a positive increase in brain weight for the mulatto, their accumulated evidence of a corresponding inferior physical development in other respects negated the benefits of miscegenation to race progress.[60]

Hunt felt that brain-weight analysis by means of autopsies confirmed the earlier pre–Civil War measurements of Samuel Morton, who measured the internal capacity of the skull in cubic inches. Since Morton's capacity for the Teutonic skull was 92 cubic inches and for the Negro 82 cubic inches, the ratio of brain weights made during the Civil War, 52.00 ounces in the white compared to 46.40 in the Negro, confirmed Morton's earlier ratio. This meant, furthermore, that the average white had a competitive advantage over the Negro of between 5.5 and 9.5 percent.[61] Hunt concluded that though the autopsy statistics were crucial, they could not determine "the ultimate capacity of the negro from that which he has thus far manifested." It meant, moreover, that autopsies of Negroes needed to be taken at intervals in the future, to determine if the effect of freedom and education led to corresponding changes in the Negro brain. Such autopsies, he felt, would resolve the controversy that existed between the environmentalists and the hereditarians.

As between the two races, the problem is: Does the large brain by its own impulses create education, civilization and refinement, or do education, civilization and refinement create the large brain? This problem might be solved by a series of researches in the weight of brain of the poor whites of the south, known as "sand hillers," "low-down people," or "crackers." With them civilization has retrograded. They came of a good stock originally, but have degenerated into an idle, ignorant and physically and mentally degraded people. Their general aspect would indicate small brains. If they are small, it is due to the absence of educational influences.[62]

With the statistical methodology of Quetelet, anthropometry and medical science ripened, bringing meaning to the vast amounts

[60] *Ibid.*, 52; Hunt, "The Negro as a Soldier," *Quarterly Journal of Psychological Medicine*, I (Oct., 1867), 175.

[61] Hunt, "The Negro as a Soldier" (1869), 52.

[62] *Ibid.*, 53.

ETHNOGRAPHICAL TABLE,

Derived from 405 Autopsies of White and Negro Brains. Made under the direction of Surgeon Ira Russell, 11th Massachusetts Volunteers.

	Number of Autopsies.	Grade of Colour.	Average weight of brain.	Maximum weight of Brain.	Minimum weight of Brain.	Brains, 60 ounces and over.	Brains, 55 and under 60 ounces.	Brains, 50 and under 55 ounces.	Brains, 45 and under 50 ounces.	Brains, 40 and under 45 ounces.	Brains, 35 and under 40 ounces.	Brains less than 35 ounces.
			oz.	oz.	oz.							
	24	White.	52-06	64	$44\frac{1}{4}$	1	4	11	7	1	...	...
	25	$\frac{3}{4}$,,	49.05	61	40	1	...	10	12	2	...	...
	47	$\frac{1}{2}$,,	47.07	57	$37\frac{3}{4}$	...	2	13	19	12	1	...
	51	$\frac{1}{4}$,	46·54	59	$38\frac{1}{2}$	...	2	10	22	11	6	...
	95	$\frac{1}{8}$,,	46·16	57	$34\frac{1}{2}$	...	1	15	50	21	7	1
	22	$\frac{1}{16}$,,	45·18	$50\frac{1}{2}$	40	...	...	3	10	9	...	...
	141	Black.	46·96	56	$35\frac{3}{4}$	...	5	42	51	38	3	...
	405	...	...	...	...	2	14	104	171	94	17	1
Autopsies of Clendenning, Sims, Reid, and Tiedemann,	278	Whites, collated from various sources,	$49\frac{1}{2}$	65	34	7	28	99	97	39	7	1

Ethnographical table showing autopsy results on white and Negro brains (from Sanford B. Hunt, "The Negro as a Soldier," *Anthropological Review* [1869]).

Courtesy of National Library of Medicine

of data accumulated during the Civil War. Medicine and anthropometry became a funding source for both military and public investigations on the varieties of man, the Negro in particular. No longer would attitudes of racial inferiority have to employ those prewar measurements and conclusions which had been tainted with proslavery arguments. Now conclusions could appear "scientific" and, indeed, "proven" on the basis of the Civil War investigations. Perhaps the greatest irony of the Civil War was that its anthropometric investigations were used in the late nineteenth century to support institutional racism. The war acted as a "carrier" for those racial attitudes that were a part of the prewar period.

POSTWAR EXAMINATIONS

One of the initial defects discovered by postwar examiners was that their evidence assumed a uniform density of the brain. This realization led to comparisons of special cerebral functions, color, and convolutions. C. Luigi Calori and A. J. Parker, for example, found that in their examination of the brains of Negroes the characteristics of the brain corresponded generally to those of the European, except in the convolutions and in the sulci, which were less marked than in the European. Both felt there was strong evidence to show that the Negro brain bore a far closer relationship to the ape.[63] J. Barnard Davis in his "Synostotic Crania among Aboriginal Races of Man" noted "the premature ossification of the sutures of the skull in arresting the full development of the brain, and so rendering it unequal to the due performance of its functions."[64]

Looking back upon the American institution of slavery, English naturalist Robert Dunn felt the Negro had lived in a constant "state of enjoyment," a state which reflected his lack of mental complexity.[65] When his toils were over, he sang, danced, and displayed

[63] C. Luigi Calori, "The Brain of a Negro of Guinea," *Anthropological Review*, VI (July, 1868), 279–85; A. J. Parker, "Cerebral Convolutions of the Negro Brain," Philadelphia Academy of Natural Sciences, *Proceedings*, XXX (1873), 11–15.

[64] Quoted in Wilson, "Brain-Weight and Size," 185.

[65] Robert Dunn, "Some Observations on the Psychological Differences Which Exist among the Typical Races of Man," Ethnological Society of London, *Transactions*, III (1863), 20.

"mild and gentle affections." His brain, in marked resemblance to the orangoutan's, developed auspiciously at first but never proceeded past that of the Caucasian in boyhood. Both gyri and sulci in the Negro brain, argued German anatomist Friedrich Tiedemann, closely resembled the structure of the brain of the orangoutan.[66] The differences in the cerebral lobes of the white and black races showed stages of evolution or planes of development. As yet, the anterior lobes of the Negro brain, which indicated "the character of his intellectual bearing," were less developed than those in the posterior, affecting his propensities and social tendencies.[67] Similarly, the American Indian had a "simplicity and regularity" of the convolutions in the frontal lobes. Like the Negro, his "nervous apparatus of the perceptive and intellectual consciousness" was far below the complexity that characterized the Caucasian. This was proof enough for Dunn to warrant the conclusion that "the large-brained European differs from, and so far surpasses the small-brained savage in the complexity of his manifestations, both intellectual and moral." The dolichocephalic with upright face, and the brachycephalic with projecting face and jaw, represented extremes in the races of mankind.[68]

More important, however, was the work of anthropologists who, in studying the infant stages of Caucasian, Negro, and anthropoid, saw signs of physiological retrogression. In the infant stage the facial features of all races, including the orangoutan, were very similar. Yet, as each developed, the physiognomy gave visible signs of change. The Negro child, for example, was born without prognathism. His facial angle as well as his coloring were closely similar to the Caucasian child. So, too, with the infant orang, whose facial features showed little resemblance to the adult orang. By puberty, however, a rapid transformation began to take effect primarily in the cranium and face. Dunn wrote, "Whilst in the white man the gradual increase of the jaws and the facial bones is not only equalled, but exceeded, by the development, or rather enlarge-

[66] *Ibid.*, 22.

[67] Robert Dunn, "Civilization and Cerebral Development; Some Observations on the Influence of Civilization upon the Development of the Brain in the Different Races of Man," Ethnological Society of London, *Transactions*, IV (1864–1865), 20.

[68] Dunn, "Some Observations," 21.

ment of the brain, and especially the anterior lobes; the reverse is the case in the Negro. The central frontal suture closes in the Negro in early youth, as well as the parietal part of the coronal suture. With advancing age the central portion of the coronal suture, the sagittal suture, and all the parietal sutures close, nearly simultaneously."[69]

While Negro and Caucasian children were equal in their infant capacities, no sooner did they reach puberty than the Negro, like the orangoutan, became incapable of further progress. With the projection of the jaws and the closure of the cranial sutures, both the Negro and the orang came to the end of their intellectual development.[70] With this as a backdrop, the intermediate Indian, Mongol, and Malay furnished similar arrested or "mummified intelligence." Though European anthropologists provided the foundation for this concept, Americans like John Fiske and Edward Drinker Cope willingly contributed to its greater elaboration. Their interest in cranial suture development, drawn in part from the English studies of Robert Dunn, Frederick W. Farrar, and the writings of Herbert Spencer and Filippo Manetta, came at the auspicious hour of America's disillusionment with Reconstruction and the popular appeal for disfranchisement.[71]

Moving from strictly cranial capacity to brain weight, suture study, disposition, convolution development, and relative proportions of the different subdivisions (cerebrum, cerebellum, pons varolii), craniometric studies in the postwar decades added more exacting standards to the test of intellectual and racial vigor.[72]

[69] Dunn, "Civilization and Cerebral Development," 25.

[70] As late as 1895 British anthropologist Augustus H. Keane referred to plantation reports in the United States regarding the arrest of intelligence in the Negro child at puberty: "The intellect seemed to become clouded, animation giving place to a sort of lethargy, briskness yielding to indolence." See his *Ethnology* (London, 1895), 266.

[71] Frederick W. Farrar, "Aptitudes of Races," Ethnological Society of London, *Transactions*, V (1866), 123; John Crawfurd, "On the Physical and Mental Characteristics of the European and Asiatic Races of Man," *ibid.*, 60; "Weight and Development of the Brain," 473; Max Bücher, "African Psychology," *Popular Science Monthly*, XXIII (July, 1883), 399–400; Ernest W. Coffin, "On the Education of Backward Races," *Pedagogical Seminary*, XV (Mar., 1908), 32–33.

[72] J. Simms, "Human Brain-Weights," *Popular Science Monthly*, XXXI (July, 1887), 355–59; C. K. Mills, "Arrested and Aberrant Development of Fissures and Gyres in the Brains of Paranoiacs, Criminals, Idiots and Negroes. Preliminary Study

While anthropologists did not take up the immediate task proposed by Sanford Hunt, they did utilize a greater number of postmortem examinations in their accumulation of evidence on the structure, size, weight, and character of the brain. One outcome of this was the creation of the Mutual Autopsy Society of Paris, founded in 1881 with the purpose of securing brains for scientific study. In the United States similar societies formed. The American Anthropometric Society, founded in 1889 by Harrison Allen, Francis Xavier Dercum, Joseph Leidy, William Pepper, and Edward Charles Spitzka, organized for the preservation of the brains of its members. The American Anthropometric Society was followed soon afterward by the Cornell Brain Association under the guidance of Professor Burt G. Wilder.[73]

One of the most extensive brain studies was performed on Major John Wesley Powell of the Bureau of Ethnology of the Smithsonian Institution following his death in 1903. Although the study evolved out of a "conversation bet" between Powell and W J McGee (1835–1912), anthropologist and ethnologist of the Bureau of Ethnology, as to who had the largest brain, the study was made in all seriousness and was carried out by brain surgeon Edward Spitzka (1876–1922).[74] Spitzka, physician and professor of general anatomy at Jefferson Medical College, directed the examination, while McGee, Frank Baker (1841–1918), professor of anatomy at Georgetown University and editor of the *American Anthropologist*, and Daniel S. Lamb (1843–1929), vice-president of the Association of American Anatomists, assisted in the examination. They analyzed Powell's brain characteristics as part of a general study of brain weights among noted people. Spitzka contributed articles to both the *American Anthropologist* and the *Philadelphia Medical Jour-*

of a Chinese Brain," *Journal of Nervous and Mental Diseases*, n.s., XI (1886), 517–53; Claude Bernard, "On the Functions of the Brain," *Popular Science Monthly*, II (Nov., 1872), 64–74.

[73] Edward C. Spitzka, "A Study of the Brains of Six Eminent Scientists and Scholars Belonging to the American Anthropometric Society, Together with a Description of the Skull of Prof. Edward D. Cope," American Philosophical Society, *Transactions*, n.s., XXI (1907), 175–76.

[74] William C. Darrah, *Powell of the Colorado* (Princeton, N.J., 1951), 390–91; Edward C. Spitzka, "A Death Mask of W J McGee," *American Anthropologist*, n.s., XV (July-Sept., 1913), 536–38.

nal on the subject of brain weights and concluded, along with corroborating studies of Johannes Ranke, Rudolph Virchow, and Leonce Manouvrier, that a definite relationship existed between cranial capacity and psychic ability. He alluded to Paul Broca's researches which showed that the skulls of modern Parisians were larger than those of twelfth-century Parisians, thus proving increased cranial development. Spitzka also looked to Dr. John Venn's article in *Nature* in 1890 which asserted that the cranial capacity of Cambridge University students was on the "average greatest and growing for the longest time in the group of the most successful men."[75]

Adding Powell's measurements to those of some 103 other brain weights, Spitzka concluded that the brain weight of noted individuals whose ages averaged 62.4 years was 1,469.65 grams, a figure which exceeded the normal European brain by 100 grams.[76] His results were "in accord with biological results as the fact that brachycephaly and increased cranial capacity in the most progressive races are in direct and intimate relation to each other." The brain-weight difference between Georges Cuvier and an African Zulu (780 grams) was in approximately the same scale, he argued, as the difference between a Zulu and a gorilla (525 grams).[77] The statistics, he argued, pointed to a gradation of the human species, "for we may have cranial capacities ranging from about 2000 cc in some of our most eminent men to less than 1000 cc in the lowly Hottentot or Florida Indian."[78]

	BRAIN WEIGHT IN GRAMS	APPROXIMATE RATIO
Turgenev	2,012	
Cuvier	1,830	1.00
General Ben Butler	1,758	
Thackeray	1,658	

[75] Edward C. Spitzka, "A Study of the Brain of the Late Major John Wesley Powell," *American Anthropologist*, n.s., V (Oct.-Dec., 1903), 591; Francis Galton, "Head Growth in the Students at the University of Cambridge," *Nature*, XXXVIII (1888), 14.

[76] Spitzka, "Study of the Brain," 600.

[77] *Ibid.*, 601–2.

[78] *Ibid.*, 603.

	BRAIN WEIGHT IN GRAMS	APPROXIMATE RATIO
Zulu	1,050	
Australian	907	0.50
Bushwoman	794	
Gorilla	425	
Orang	400	0.25
Chimpanzee	390	

Anthropometrical characteristics other than the head were not given much notice in the early years of race study. Generally, with the notable exceptions of the Hottentot Venus and pygmy, the plurality of racial types was given full expression in the facial attitude rather than in osteometrical peculiarities. The importance of bodily proportions—the perforation of the humerus, the curvature of the femur, the angle which the body makes with the diaphysis, the angle of torsion of the humerus—gained credence principally with the acceptance of race evolution after Darwin. Anthropometric differences had been observed by Jeffries Wyman, Joseph Leidy, Paul Broca, J. Barnard Davis, E. T. Hamy, William Lawrence, and Pruner-Bey, but such osteological studies did not become a part of the comparative study of races until physicians were permitted to make large-scale studies of men of similar age and postmortem studies in the dissecting rooms. There, along with measuring the proportions of the skeleton, physicians began to study muscles, viscera, vessels, and nerves for comparative analysis.[79] Ironically, however, it was the southern physician who generally carried out studies on the Negro in the late nineteenth century, and his conclusions reflected not only the section's appeal for a reappraisal of Reconstruction politics but also mirrored the race ideology of the antebellum South.

[79] Jeffries Wyman, "Observations on the Skeleton of a Hottentot," Boston Society of Natural History, *Proceedings*, IX (1865), 352–57; A. Hunter Dupree, "Jeffries Wyman's Views on Evolution," *Isis*, XLIV (1953), 243–46; G. D. Gibbs, "Essential Points of Difference between the Larynx of the Negro and the White Man," Anthropological Society of London, *Memoirs*, II (1865), 1–13; Joseph Leidy, "A Lecture on the Anatomical Peculiarities of the Negro," *Medical and Surgical Reporter*, X (1857), 228; W. E. Horner, "On the Odoriferous Glands of the Negro," *American Journal of Medical Science*, n.s., XI (1846), 13–16.

II *The Physician versus the Negro*

THE STATISTICIAN and superintendent of the Eighth Census, Joseph Camp Kennedy, remarked in 1862 that the gradual extinction of the Negro race was an "unerring certainty." He suspected, furthermore, that the freedom of the Negro race—and its integration into a white society—would only hasten the process.[1] Kennedy's view, further confirmed by a faulty statistical analysis in the Ninth and Tenth Censuses of 1870 and 1880, the corroborating beliefs of physicians, the investigations of American insurance companies, the statistical evidence of the United States Army, as well as countless medical reports, precipitated a belief in the Negro's inevitable extinction.[2] Even Major General O. O. How-

[1] J. Stahl Patterson, "Increase and Movement of the Colored Population," *Popular Science Monthly*, XIX (Sept., 1881), 667; Joseph Camp Kennedy, *Preliminary Report on the Eighth Census* (Washington, D.C., 1862), 8.

[2] Patterson, "Increase and Movement," 667–68; U.S. Bureau of the Census, *Negro Population, 1790–1915* (Washington, D.C., 1918), 18; Henry Gannett, "Was the Count of Population in 1890 Reasonably Correct?" American Statistical Association, *Publications*, IV (1895), 99–102; Gannett, "Statistics of the Negroes in the United States," John F. Slater Fund, *Occasional Papers*, no. 4 (1894), 24; Francis A. Walker, "Statistics of the Colored Race in the United States," American

ard of the Freedmen's Bureau was concerned enough to send J. W. Alvord, his general superintendent of education, on a trip through the southern states to verify the growing beliefs that Negroes were "diseased and degraded," that "they [were] all dying off," that "they [were] killing their children," and that "they [would] not work."[3] To be sure, there were skeptics who pointed to the absolute numerical increase of the Negro in America, as opposed to the apparent percentage rate decline of Negroes in the total population.[4] There were also those elements in American society for whom the wish for the Negro's extinction "was father to the thought." For them, the belief usually required a need for continuous reassertion that the evidence was, indeed, truthful.[5] Yet, despite the complexity of the problem and the reservations of many, the belief in the Negro's extinction became one of the most pervasive ideas in American medical and anthropological thought during the late nineteenth century. It was also a fitting culmination to the concept of racial inferiority in American life.

The census reports, along with insurance and army statistics on the apparent decline in vitality of the American Negro, brought

---

Statistical Association, *Publications*, II (1890), 91–106; Samuel J. Holmes, *The Negro's Struggle for Survival* (Berkeley, Calif., 1937), 14–16; George W. Williams, *History of the Negro Race in America from 1619 to 1880*, 2 vols. (New York, 1883), II, 417, 549–51; U.S. Department of Labor, "Condition of the Negro in Various Cities," *Bulletin*, no. 10 (1897), 257–369; W. J. Trent, Jr., "Development of Negro Insurance Enterprises" (M.B.A. thesis, University of Pennsylvania, 1932), 17–18; Winfred O. Bryson, Jr., "Negro Life Insurance Companies" (Ph.D. thesis, University of Pennsylvania, 1948), 7, 46, 99, 30, 26; "The Colored Race in Life Assurance," *Lancet* (London), II (Oct., 1898), 902; Frederick L. Hoffman, *History of the Prudential Insurance Company of America* (Newark, N.J., 1900), 153, 210–11; Hoffman, "The Statistical Experience Data of the Johns Hopkins Hospital, 1892–1911," Johns Hopkins Hospital, *Reports*, XVII (1916), 185–345.

[3] J. W. Alvord, *Letters from the South Relating to the Condition of the Freedmen Addressed to Major General O. O. Howard* (Washington, D.C., 1870), i, 23.

[4] E. W. Gilliam, "The African in the United States," *Popular Science Monthly*, XXII (Feb., 1883), 433–45; Williams, *History of the Negro Race in America*, II, 417–18; Holmes, *The Negro's Struggle for Survival*, 1–5; Carter G. Woodson, *A Century of Negro Migration* (Washington, D.C., 1918), chap. I; U.S. Bureau of the Census, *A Century of Population Growth from the First Census of the United States to the Twelfth, 1790–1900* (Baltimore, 1967), 92; Philip A. Bruce, *The Plantation Negro as a Freeman: Observations on His Character, Condition, and Prospects in Virginia* (New York, 1889), 261–62.

[5] Williams, *History of the Negro Race in America*, II, 417; Henry Gannett, "Are We to Become Africanized?" *Popular Science Monthly*, XXVII (June, 1885), 145–50.

the responsibility squarely upon the physician in America to continue to study his physical and mental makeup and to follow his course of health through the postwar years. The United States Sanitary Commission, along with the United States Surgeon-General's Office and the Provost Marshal–General's report of the United States Army, had given the initial warning of the deterioration occurring in the Negro race.[6] Following this alarm, physicians, mostly from the South, began to study and publish reports on the comparative mortality, health, and physiognomy of the two races. Their studies suggested that a fundamental change was taking place in the physiological and pathological makeup of the Negro since the days of slavery—the postwar Negro was succumbing to disease in far greater numbers than the antebellum generation, and the Negro's future seemed precariously close to extinction. The same situation which placed the southern physician in the position of analyzing the Negro's health status also permitted him to elaborate on the reasons behind the apparent problems. Physicians, given the responsibility for presenting a clinical analysis, roamed far and wide in their studies and conclusions.

Census statistics were the basis for much of the speculation. The Ninth Census (1870) had shown that the white population during the years from 1860 to 1870 had increased 34.76 percent while blacks had increased a mere 9.86 percent. This news came as a surprise, since the previous rates of increase for blacks in America had averaged 29.98 percent in the census reports from 1790 to 1850. These new statistics caused immediate concern and speculation as to the future of the Negro race. The Ninth Census report was offset by the Tenth, which showed a comparative rate of increase

[6] Benjamin A. Gould, *Investigations in the Military and Anthropological Statistics of American Soldiers* (New York, 1869); J. H. Baxter, *Statistics, Medical and Anthropological, of the Provost Marshal–General's Bureau*, 2 vols. (Washington, D.C., 1875); U.S. Surgeon-General's Office, *The Medical and Surgical History of the War of the Rebellion*, 3 vols. (Washington, D.C., 1870–1888), I, pt. 3; Joseph R. Smith, "Sickness and Mortality in the Army," American Medical Association, *Transactions*, XXXIII (1882), 311–15; Robert Reyburn, "Type of Disease among the Freed People (Mixed Negro Races) of the United States," *Medical News* (Philadelphia), LXIII (Dec., 1893), 623–27; Sanford B. Hunt, "The Negro as a Soldier," *Quarterly Journal of Psychological Medicine*, I (Oct., 1867), 161–87; "Editorial," *New York Medical Times*, XXIII (Mar., 1895), 93–94; Henry Latham, *Black and White* (London, 1867), 275–76.

from 1870 to 1880 of 29.22 percent for the Caucasian and 34.85 percent for the Negro. However, the Eleventh Census again reversed the black increase. From 1880 to 1890 whites increased 26.68 percent while blacks increased only 13.53 percent. The census officers, looking back on the 100 years of census statistics, remarked that "the whites increased from 80.83 to 87.80, while the colored element today is two-thirds less than it was a hundred years ago."[7] Statistics from Missouri, Texas, Alabama, Maryland, Georgia, North Carolina, South Carolina, Louisiana, Tennessee, Virginia, and the District of Columbia suggested that while emancipation had begun the Negro's "career as a freedman and the struggle for elevation," it also had led to his "physical decline."[8] In Charleston, South Carolina, for example, both the white and the black man in 1860 had a mortality rate of 12 per 1,000, but by 1895 the Negro mortality rate had climbed to 29.1 per 1,000 while the white death rate had increased to 18.7 per 1,000.[9] Dr. John H. Van Evrie of New York suggested that the Negro's tendency toward race extinction "accelerated or diminished in exact proportion as 'impartial freedom' [was] thrust upon him." As the Negro began to enjoy equality as a result of the "blind and cruel kindness and exterminating goodness" of the Caucasian, he succumbed to the harshness of natural race laws.[10]

During the 1880's and 1890's medical doctors initiated newer studies on the Negro in an effort to analyze the effect of freedom upon his physical, mental, and moral capacity. They compared and contrasted their evidence with the medical investigations made by both the army and the United States Sanitary Commission during the Civil War. Doctor Thomas P. Atkinson of Virginia reminded

[7] J. T. Walton, "The Comparative Mortality of the White and Colored Races in the South," *Charlotte Medical Journal*, X (1897), 292; S. S. Herrick, "Comparative Vital Movement of the White and Colored Races in the United States," *New Orleans Medical and Surgical Journal*, n.s., IX (1881–1883), 678.

[8] Walton, "Comparative Mortality," 292; A. R. Kilpatrick, "An Account of the Colored Population of Grimes County, Texas. A Comparison between Their Present and Former Condition," *Richmond and Louisville Medical Journal*, XIV (Nov., 1872), 610; L. S. Joynes, "Remarks on the Comparative Mortality of the White and Colored Population of Richmond," *Virginia Medical Monthly*, II (June, 1875), 155.

[9] "The Negro," *Atlanta Journal-Record of Medicine*, V (June, 1903), 186–87.

[10] John H. Van Evrie, *White Supremacy and Negro Subordination* (New York, 1868), 311–12.

the medical profession that there was a much higher death rate among black soldiers than among whites during the war and that this rate continued after the war. From the wartime evidence of Baxter's medical and anthropometrical history and from his own statistics, Atkinson concluded that not only had the Negro deteriorated mentally, morally, and physically from his earlier condition in slavery, but "a different mode of treatment is indicated in the management of his diseases."[11] The secretary of the Texas State Medical Association, Dr. W. J. Burt, also drew upon Civil War medical reports to confirm his own belief in Negro inferiority. He concluded that not only did osteological measurements made during the war place the Negro "next below man in the zoological scale," but the Negro's physiological peculiarities made him more susceptible to disease and death.[12] Burt relied upon the postmortem examinations of Dr. A. McDowell on white and colored troops, which discovered differences in chest measurements, lung weight, and size of liver and spleen.[13] From these bodily differences and the fact that the Negro brain was about one-eighth less than that of the Caucasian, Burt argued that the Negro seldom endured surgical operations due to his lack of "nervous endurance and fortitude." The Negro's physiological inferiority, his poorer "mental manifestation and power," and his lack of "moral courage" made him unable to withstand surgical operations.[14]

From the wartime statistics of Drs. George A. Otis and Joseph J. Woodward on Negro mortality, other physicians asserted that the sudden susceptibility of the emancipated Negro to disease demonstrated the consequences of breaking his natural race path-

[11] Thomas P. Atkinson, "On the Anatomical, Physiological and Pathological Differences between the White and the Black Races," Medical Society of Virginia, *Transactions* (1873), 67; Smith, "Sickness and Mortality in the Army," 313–14; Reyburn, "Type of Disease," 624, 626.

[12] W. J. Burt, "Report on the Anatomical and Physiological Differences between the White and Negro Races, and the Modification of Disease Resulting Therefrom," *St. Louis Courier of Medicine*, VIII (Nov., 1882), 419; Burt, "Report on the Anatomical and Physiological Differences between the White and Negro Races," Texas Medical Association, *Transactions*, VIII (1876), 115–23.

[13] Burt, "Report" (1882), 420.

[14] *Ibid.*, 421; discussion in Louis McLane Tiffany, "Comparison between the Surgical Diseases of the White and Colored Races," American Surgical Association, *Transactions*, V (1887), 272.

ology which had existed within the framework of the institution of slavery.

> Surely there must be something more than mere chance in this sudden reversion of settled facts. Was there not something in the rigid regime under which the slave lived that rendered his system a barren soil to the germs of tuberculosis? . . . and this change has come, in my opinion, as the result of the violent striking of the shackles from the hands of a people who, for generations, had lived as slaves; the sudden lifting of all restraint, the violent swing of the pendulum from a simple life of toil and bondage to one of liberty, license, and all that inevitable brood of disasters that follows surely and swiftly upon the heels of outraged and violated natural laws.[15]

Dr. J. F. Miller, superintendent of Eastern Hospital in Goldsboro, North Carolina, published a study in the *North Carolina Medical Record* for 1896 which attempted to judge the effects of emancipation upon the mental and physical capacities of the Negro. Using the statistics of the superintendent of the Georgia lunatic asylum, he recalled that the number of insane Negroes had increased measurably since emancipation. While in 1860 there were but forty-four insane Negroes in the state of Georgia, or one in every 10,584 of the population, the censuses of 1870, 1880, and 1890 showed significant increases. The census of 1890 showed an increase of insane Negroes to one in every 943 of the population.[16] The untutored slave, wrote Miller, with "no thought for the morrow, wherewithal he should be fed and clothed," had no ambitions, hopes, or possibilities in his future. His quiet "humble life in his little log cabin, with his master to care for every want of self and family, in sickness and in health," had been commensurate with his physiological and mental condition.[17] The violation of those

[15] F. Tipton, "The Negro Problem from a Medical Standpoint," *New York Medical Journal*, XLIII (May, 1886), 570; J. G. Rogers, "The Effect of Freedom upon the Physical and Psychological Development of the Negro," American Medico-Psychological Association, *Proceedings*, VII (1900), 88–99.

[16] J. F. Miller, "The Effects of Emancipation upon the Mental and Physical Qualifications of the Negro in the South," *North Carolina Medical Journal*, XXXVIII (Nov., 1896), 287; T. O. Powell, "The Increase of Insanity and Tuberculosis in the Southern Negro Since 1860, and Its Alliance and Some of the Supposed Causes," American Medical Association, *Journal*, XXVII (1896), 1185–88.

[17] Miller, "Effects of Emancipation," 289.

natural laws by emancipation had "left its slimy trail of sometimes ineradicable disease upon [the Negro's] physical being," and the licentiousness Miller thought evident in the freed Negro brought upon him "a beautiful harvest of mental and physical degeneration and he is now becoming a martyr to an heredity thus established." While the Negro could live in comfort "under less favorable circumstances than the white man, having a nervous organization less sensitive to his environments," Miller wrote, "yet it is true that he has less mental equipoise, and may suffer mental alienation from influences and agencies which would not affect a race mentally stronger."[18]

Without a proper ancestry conditioned by the responsibilities of freedom and without the education or preparedness for responsibility, the Negro citizen, thrust into a modern world which he had in no way helped to create, deteriorated under the strain.[19] French anthropologist Paul Topinard noted a perceptible increase in the relative frequency of mania and idiocy in the Negro population after emancipation. Forced to do "battle with the necessities of the social condition" of freedom, the emancipated slave fell victim to the vicissitudes of mental disorder.[20] So overwhelming was the strain, wrote Dr. J. Allison Hodges, dean of the College of Medicine in Richmond, that the Negro in America was either dying out or reverting to his primitive savagery, as evidenced in his unrestrained sexual passion.[21] It seemed clear to many physicians that "the American negro [would] never become firmly established in the right methods of living before disease and death . . . thinned his ranks and there will be no race problem."[22] Even if, as some speculated, the Negro population was indeed increasing, physicians argued that such increase could only come at the expense of moral and physical development. "When the pendulum swings the other way," wrote a member of the American Medical

[18] *Ibid.*, 290.

[19] *Ibid.*, 292–93.

[20] Paul Topinard, *Anthropology* (London, 1878), 413.

[21] J. Allison Hodges, "The Effect of Freedom upon the Physical and Psychological Development of the Negro," *Richmond Journal of Practice*, XIV (June, 1900), 170–71.

[22] Seale Harris, "The Future of the Negro from the Standpoint of the Southern Physician," *Alabama Medical Journal*, XIV (Jan., 1902), 58.

Association, "the increase in the defective classes will only hasten the final destruction of the race."[23] The Atlanta National Conference of Charities and Corrections in 1903 warned that the Negro race was "in danger of being destroyed by insanity."[24]

"That the immediate emancipation of the Southern negro was a most deplorable event in the history of that unhappy race," wrote Dr. E. T. Easley of Dallas, Texas, "has become quite manifest." The new status of freedom brought upon the Negro the full effects of race struggle and consequent race deterioration. Those who "knew him best and were most familiar with his habits of life and constitution" were now his caretakers or, more correctly, his undertakers. His present status in the body politic had not only proved the southern argument that the Negro was "notoriously incompetent" but also that he could not exist as an equal in a free society.[25] Dr. Eugene R. Corson of Savannah, Georgia, in an article in the *New York Medical Times* of 1887, took issue with men like sociologist E. W. Gilliam and jurist Albion Tourgée, who prophesied a steady increase of the black population in America. Corson accepted the fact that the colored race was more prolific than the white, but he argued that it was only in "potential prolificness"—the explanation which Herbert Spencer had resolved in his *Principles of Biology*. "The simpler the organism," Corson wrote, "the simpler the genesis and the greater the prolificness." The white race, with its greater mental organization and differentiation, had a "more complex . . . genesis and . . . less . . . prolificness." But though prolificness was small with the Caucasian, "it is more than compensated for by the ability to maintain individual life."[26] The Negro, on the other hand, forcibly transferred to America from his natural habitat, did not live under conditions most favorable for

[23] *Ibid.*, 62.

[24] "Deterioration of the American Negro," *Atlanta Journal-Record of Medicine*, V (July, 1903), 287; H. M. Bannister and Ludwig Hektoen, "Race and Insanity," *American Journal of Insanity*, XLIV (Apr., 1888), 63.

[25] E. T. Easley, "The Sanitary Condition of the Negro," *American Medical Weekly*, III (July, 1875), 49.

[26] Eugene R. Corson, "The Future of the Colored Race in the United States from an Ethnic and Medical Standpoint," pt. I, *New York Medical Times*, XV (Oct., 1887), 193, 196; Herbert Spencer, "A Theory of Population Deduced from the General Law of Animal Fertility," *Westminster Review*, LVII (1852), 468–501.

his racial development. Thrown into "the struggle for existence" with a civilization "of which he is not the product," the Negro "must suffer physically, a result which forbids any undue increase of the race, as well as the preservation of the race characteristics." Corson felt it was possible to predict the future of the Negro race; its inferior race status, along with its forcible transportation to America and the later emancipation from slavery, were bringing to issue evident signs of deterioration and degeneration. Though Malthus had thrown light upon the laws governing population increase, it was "to the school of Darwin, Wallace, and Spencer that we must turn for valuable teachings, and the elucidation of laws which constitute all potent factors in the growth and development of humanity."[27]

Dr. R. M. Cunningham, a former penitentiary physician from Alabama, reminded the profession that just as there were innate hereditary influences which prompted the Negro to acts of crime, so there were also anatomical and physiological differences between him and the Caucasian—differences which made him not only inferior to the white man but which predisposed him to disease, high mortality, and race deterioration. Cunningham's findings, along with those of Colonel John G. Milner on the New Jersey, New York, Pennsylvania, and Ohio prison systems, presented statistical evidence of the Negro's higher prison mortality rate. The real significance of their data on the Negro, argued Cunningham, stemmed from the fact that the prison system provided the same environmental factors to both blacks and whites. And since the same had been true of the Negro and white soldier during the years of the Civil War—an environment in which both races endured the same hardships, hygienic life, and nursing facilities—Cunningham felt the evidence strongly pointed to inherent race deficiencies. Even after the war army medical reports showed the annual death rate of the Negro to be more than double that of the white soldier, a situation comparable to the statistical evidence in the prison system.[28]

[27] Corson, "Future of Colored Race," 197–98.

[28] R. M. Cunningham, "The Morbidity and Mortality of Negro Convicts," *Medical News* (Philadelphia), LXIV (Feb., 1894), 113, 116; Sanford B. Hunt, "The

That there were major anatomical differences between the Negro and the white was commonly accepted by physicians throughout the late nineteenth century. The Association of American Anatomists circulated a questionnaire that asked physicians to "keep a careful record of all variations and anomalies" between the two races.[29] Osteological peculiarities had been observed by anthropologists before the nineteenth century, but generally racial peculiarities focused upon facial characteristics. What observations that existed were often isolated experiments using a variety of measuring instruments and criteria. A study by Dr. D. Kerfoot Shute of Washington, D.C., on osseous structures concluded that by an examination of anatomical peculiarities, it was possible to "stamp a race as high or low." The evolution of races, developing from the anthropoids through the savage tribes and finally culminating in the advanced civilizations, showed the development of an upright posture which led to corresponding changes in the thorax, pelvis, and lumbar vertebrae. Just as the skull became less prognathous as the races began their slow ascent to higher intellectual attainment, so the posture "shifted the weight of the abdominal viscera from the thorax to the pelvis . . . and also the last lumbar vertebra tend[ed] to fuse with the sacrum, thus tilting up still further the pelvis."[30] Similarly, Dr. Van Evrie felt that because of the Negro's physiological place in nature, he was "incapable of an erect or direct perpendicular posture." The structure of his limbs, the form of pelvis and spine, and the way the head was set on the shoulders gave the Negro a "slightly stooping posture"—a position in evolution that was nearer the anthropoid than the Caucasian.[31] Characteristics that were simian—flattened tibia, narrow pelvis, elongated calcaneum, long and perforated humerus

---

Negro as a Soldier," *Anthropological Review*, VII (Jan., 1869), 42–43; "Editorial," *New York Medical Times*, XXIII (Mar., 1895), 93; Atkinson, "Anatomical, Physiological and Pathological Differences," 68.

[29] Edward A. Balloch, "The Relative Frequency of Fibroid Processes in the Dark-Skinned Races," *Medical News* (Philadelphia), LXIV (Jan., 1894), 29.

[30] D. Kerfoot Shute, "Racial Anatomical Peculiarities," *American Anthropologist*, o.s., IX (Apr., 1896), 124, 126; "Racial Anatomical Peculiarities," *New York Medical Journal*, LXIII (Apr., 1896), 500–501; Arthur de Gobineau, *The Inequality of the Human Races* (New York, 1915), 114–15.

[31] Van Evrie, *White Supremacy and Negro Subordination*, 93.

—relegated the Negro to the bottom of the scale of race development. Osteometrical differences of body linearity, as well as internal anatomical differences, corroborated such skull peculiarities as wide nasal aperture, ankylosed nasal bones, prognathism, receding chin, and well-developed wisdom teeth and created an index for a hierarchy of the races.[32]

Throughout the late nineteenth century the physician remained the chief source of information for comparative race analysis. Dr. Edward A. Balloch, an instructor of minor surgery in the medical department of Howard University, felt secure in establishing several anthropometric generalities concerning the Negro race.

> The skull is prognathous, with a facial angle of from 65 to 70 degrees. The parietal bones are thick. The zygomatic arches are wide, and the upper edge of the orbit projecting. The pelvis is long and narrow, the iliac bones less wide and more vertical. The tibia and fibula are more convex, and the os calcis is continued in a straight line with the other bones of the foot. The scapulae are shorter and broader. The thigh and arm are rather shorter. While the leg is actually about the same in length, it is relatively smaller, owing to less average stature. The forearm is longer, both actually and relatively. The foot is an eighth, and the hand a twelfth, longer than in Europeans.[33]

The Negro body, according to Dr. William T. English of Pittsburgh, Pennsylvania, "present[ed] a coarseness or rudeness and a variety in symmetry with other mathematical inaccuracies." What this meant in anthropometric terms was a heavy, thick, and coarse skull, bones which when examined microscopically "show[ed]

[32] J. Arthur Thomson, "The Influence of Posture on the Form of the Articular Surfaces of the Tibia and Astragalus in the Different Races of Man and the Higher Apes," *Journal of Anatomy and Physiology*, XXIII (July, 1899), 616–39; "Parturition of the Negro," *Philadelphia Medical Times*, XVI (1885), 296; S. M. Burnett, "Racial Influence in the Etiology of Trachoma," *Medical News* (Philadelphia), LVII (1890), 542; R. M. Fletcher, Jr., "Surgical Peculiarities of the Negro," Medical Association of Alabama, *Transactions* (1898), 49–57; A. H. Frieberg and J. H. Schroeder, "A Note on the Foot of the American Negro," *American Journal of Medical Science*, CXXVI (1903), 1033–36; J. T. Johnson, "On Some of the Apparent Peculiarities of Parturition in the Negro Race, with Remarks on Race Pelvis in General," *American Journal of Obstetrics*, VIII (1875–1876), 88–123; A. J. Parker, "Simian Characteristics in Negro Brains," Philadelphia Academy of Natural Sciences, *Proceedings*, XXXI (1879), 339.

[33] Balloch, "Relative Frequency of Fibroid Processes," 30.

comparatively less minute capillary distribution," and hands and feet which, corresponding to the Negro's evolution, marked him as a race "come out of the depths of centuries." Such peculiarities, brought to a climax with the political immediacy of the Civil War and emancipation, doomed the race to high mortality. Paget's disease, rickets, hip-joint disease, giantism, exophthalmic goiter, thymic angiomatoses, and other diseases cut sharp inroads into the future of the race. Respiratory organs, vulnerable in their lack of stamina, yielded to pneumonia, pleurisy, pulmonary tuberculosis, and associated lung diseases which all but settled the race problem by way of outright elimination.[34]

There was a certain morbidness in the physician's emphasis upon the sexual appetite in the Negro race. The greater abdominal and genital development of the Negro merely corroborated the inferiority of his other anatomical peculiarities—his black skin, flat nose, lesser cranial and thoracic development.[35] The Negro's lower level of consciousness left him out of touch with the higher forms of human experience and weaved a corporeal structure, almost vestigial in nature, whose sexual characteristics reflected those "sexual extremes [which] belong to the age of awakening consciousness, or nascent intelligence, a stage of incipiency to moral and mental development."[36] The Negro brain, some one thousand years "behind . . . the white man's brain in its evolutionary data," existed within a visceral and organic structure that was physiologically juxtaposed to its intellectual capacity.[37] The Negro's "moral delinquencies," along with elements of "bestiality and gratification," were demonstrations of the close relationship of the race to its "animal subhuman ancestors."[38] Confined within narrow physical functions, the Negro's nearness to a superior race merely accelerated his "innate tendency to sex appetite."[39]

Physicians often emphasized the extreme precocity and early

[34] William T. English, "The Negro Problem from the Physician's Point of View," *Atlanta Journal-Record of Medicine*, V (Oct., 1903), 462–63; Harris, "Future of the Negro," 58, 60.

[35] Cunningham, "Morbidity and Mortality," 115.

[36] English, "The Negro Problem," 467.

[37] *Ibid.*, 463.

[38] *Ibid.*, 468.

[39] *Ibid.*, 465.

mental arrest of Negro youths. Growing to maturity much faster than white children, Negroes exhibited sexual passion at an earlier age and then, because of mental atrophy, remained through life seemingly enslaved to the sexual impulse. "The conflict for existence between brain growth and reproductive organ growth at puberty," wrote Dr. Eugene S. Talbot, resulted for both full black and mulatto "in the triumph of the reproductive," a situation which by itself ruled against miscegenation as a race solution.[40] "The premature closing of cranial sutures and lateral pressure of the frontal bone," stated the *Encyclopaedia Britannica* (ninth edition), necessarily limited Negro development to the lower functions of life.[41] Morality was a joke among Negro society, wrote Dr. Thomas W. Murrell of Richmond, Virginia. Morality was "assumed as a matter of convenience or when there is a lack of desire and opportunity." The increase of births in the years immediately after the Civil War was the result of the sexual license brought on by emancipation and the degeneration of marriage into "a plaything of sexual impulse."[42] Sexual license and the accompanying spread of venereal disease resulted in large numbers of Negro stillbirths and infant mortality in the 1880's and 1890's. In Charleston, wrote Dr. Easley, there were 147 Negro stillbirths to 26 white. "Besides," added Dr. Lebby, the registrar of Charleston, "there were the large number put away in vaults, gardens, and rivers."[43]

Statistics of stillbirths among blacks caused many doctors to suggest that syphilis was one of the causative factors in the mortality of the race.[44] According to some physicians, the apparent increase in black population in the immediate years after the Civil War was evidence of the promiscuity of the freed Negro. But "nature abhors promiscuous sexual intercourse," wrote Dr. Seale

40 Eugene S. Talbot, *Degeneracy, Its Causes, Signs, and Results* (London, 1899), 102.

41 Quoted in C. C. Mapes, "Remarks from the Standpoint of Sociology," *Medical Age* (Detroit), XIV (1896), 714.

42 Thomas W. Murrell, "Syphilis and the American Negro—a Medico-Sociological Study," Medical Society of Virginia, *Transactions* (1909), 169.

43 Easley, "Sanitary Condition of the Negro," 49.

44 Harris, "Future of the Negro," 63; Hunter McGuire and G. Frank Lydston, "Sexual Crimes among the Southern Negroes; Scientifically Considered," *Virginia Medical Monthly,* XX (May, 1893), 106.

Harris, vice-president of the Tri-State Medical Society of Georgia, "and the abuse of the organs of reproduction will certainly result in their becoming functionless."[45] The Ninth Census report seemed to confirm his suspicions of the irreparable damage done to the sex organs. "What can we expect of the negro," he wrote, "but that he will in time share the fate of the North American Indian."[46]

Dr. G. Frank Lydston, professor of genito-urinary surgery and syphilology of the Chicago College of Physicians, considering the rape of white women by Negroes, argued that the phenomenon was related to the Negro's particular development in the evolutionary schema which "can be only said to have fairly begun with his liberation."[47] When a race "of a low type of development is subjected to an emotionally intellectual strain," its primitive instincts are brought to the surface in manifestations of lust or "bloodthirstiness, singly or combined."[48]

> When all inhibitions of a high order have been removed by sexual excitement, I fail to see any difference from a physical standpoint between the sexual furor of the negro and that which prevails among the lower animals in certain instances and at certain periods . . . namely, that the *furor sexualis* in the negro resembles similar sexual attacks in the bull and elephant, and the running amuck of the Malay race. This *furor sexualis* has been especially frequent among the negroes in States cursed by carpetbag statesmanship, in which frequent changes in the social and commercial status of the negro race have occurred.[49]

Along with the emphasis upon sexual passion, there was an equal emphasis given to the development of the "virile organs," which often reached "massive proportions." That the Negro penis exceeded in size that of the average adult white male was universally accepted as true. The Negro woman was also marked with sexual differences. Speculation generally concerned the posi-

[45] Harris, "Future of the Negro," 62; Frederick L. Hoffman, "Vital Statistics of the Negro," *Arena*, XXIX (Apr., 1892), 534.

[46] Harris, "Future of the Negro," 65.

[47] McGuire and Lydston, "Sexual Crimes," 110; "As Ye Sow That Ye Also Reap," *Atlanta Journal-Record of Medicine*, I (June, 1899), 266–67.

[48] McGuire and Lydston, "Sexual Crimes," 111.

[49] *Ibid.*, 118; Daniel G. Brinton, *The Basis of Social Relations* (New York, 1902), 160–61.

tion of the hymen, early menstruation, and the frequent "atrophic condition of the external genital organs in which the labia are much flattened and thinned, approaching in type that offered by the female anthropoid ape, hepale [sic], lemur and other pithecoid animals."[50] The sexual characteristics of the Negro male and female, their "utter contempt and cynical disbelief in the existence of chastity," as well as the male's "stallion-like passion and entire willingness to run any risk and brave any peril for the gratification of his frenetic lust," made the Negro a menace to the Caucasian race.[51]

William Lee Howard, a physician from Baltimore, suggested atavism in the Negro race in a scurrilous attack published in *Medicine* in 1903. The truth is, he said, that the Negro was "returning to a state of savagery." His "sexual madness" and religious emotionalism were marks of the "innate character of the African." Freedom for the Negro meant a return to his "ancestral sexual impulses." By understanding the anatomical and physiological character of the African, Howard thought he could "scientifically and humanely place" the Negro in the biological scale of nature far below the Caucasian. That "a few years of Latin and Bible teaching" could restrain his "periodical erethisms of the sexual centers" was wholly unwarranted. Neither education nor religious training could change "the anatomical and physiological reason for his sexuality and bestiality." Using the researches of Scottish biologists Patrick Geddes and J. Arthur Thomson, authors of studies ranging from city planning to evolution and sex, the Baltimore physician noted that whether the Negro characteristics were exaggerated or

[50] "Genital Peculiarities of the Negro," *Atlanta Journal-Record of Medicine,* IV (Mar., 1903), 842, 844; Mapes, "Remarks," 713; E. B. Turnipseed, "Hymen of the Negro Women," *Richmond and Louisville Medical Journal,* VI (1868), 194–95; Turnipseed, "Some Facts in Regard to the Anatomical Difference between the Negro and White Races," *American Journal of Obstetrics and Diseases of Women and Children,* X (1877), 33; William Lawrence, *Lectures on the Comparative Anatomy and the Natural History of Man* (London, 1840), 285; C. H. Fort, "Some Corroborative Facts in Regard to the Anatomical Difference between the Negro and White Races," *American Journal of Obstetrics and Diseases of Women and Children,* X (1877), 258–59.

[51] "Genital Peculiarities of the Negro," 844; William H. Holcombe, "Characteristics and Capabilities of the Negro Race," *Southern Literary Messenger,* XXXIII (Dec., 1861), 401–10.

lessened, one could not obliterate them. "What was decided among prehistoric Protozoa," he wrote, quoting Geddes and Thomson, "cannot be changed by act of Congress."[52] The Negro's attack on the Caucasian woman was a radical instinct as impossible to change as "the inherent order of the race."

It is this sexual question that is the barrier which keeps the philanthropist and moralist from realizing that the phylogenies of the Caucasian and African races are divergent, almost antithetical, and that it is gross folly to attempt to educate both on the same basis. When education will reduce the large size of the negro's penis as well as bring about the sensitiveness of the terminal fibers which exist in the Caucasian, then will it also be able to prevent the African's birthright to sexual madness and excess—from the Caucasian's viewpoint.[53]

The late nineteenth-century emphasis on the Negro's sexual organs, of course, was nothing new to medical or anthropological study. It was a curiosity which had a long history, beginning with the earliest European contact with the African. Ostensibly, the examination of sexual characteristics was part of a general inquiry into the comparative sexual anatomy of orang, African, and Caucasian in an effort to ascertain whether the lower races of man were closer structurally to the anthropoid or to the higher races. Hence, the groundwork for a later anti-miscegenation policy was part of a much earlier anthropological curiosity. By determining the direction of the vagina, the position of the hymen, and the general structure of the sexual organs, white doctors could set the African apart as a distinct and inferior species of man.[54] The conclusion generally drawn from sexual differentiation was that while sexual intercourse between the Caucasian male and Negro female was possible and fertile, it was "unnatural" and unproductive between the Negro male and the Caucasian female.

[52] William Lee Howard, "The Negro as a Distinct Ethnic Factor in Civilization," *Medicine* (Detroit), IX (June, 1903), 423; Patrick Geddes and J. Arthur Thomson, *Evolution* (New York, 1911).

[53] Howard, "The Negro," 424.

[54] Sir John Barrow, *Travels into the Interior of Southern Africa . . .* (London, 1806), 278–79; Jeffries Wyman and Thomas S. Savage, "Troglodytes Niger," *Boston Journal of Natural History*, IV (1843), 19; Thomas Bendyshe, ed., *The Anthropological Treatises of Johann Friedrich Blumenbach* (London, 1865), 169, 384.

One of the characters of the Ethiopian race consists in the length of the penis compared with that of the Caucasian race. This dimension coincides with the length of the uterine canal in the Ethiopian female, and both have their cause in the form of the pelvis in the Negro race. There results from this physical disposition, that the union of the Caucasian man with an Ethiopian woman is easy and without any inconvenience for the latter. The case is different in the union of the Ethiopian with a Caucasian woman, who suffers in the act, the neck of the uterus is pressed against the sacrum, so that the act of reproduction is not merely painful, but frequently non-productive.[55]

The reputed perversion of the Negro that prompted him to attack white women threatened future Caucasian evolution. "Self-constituted philosophers," who had endeavored to bring the Negro over the centuries into white society as an equal, lost sight of the difference in his "sex-diathesis." "To extend to a race, through false teaching, an egotism which should only be acquired through gradual evolvement from rational humility, is criminality to that race," wrote Pittsburgh physician English. Above all else, American society needed to preserve the Caucasian woman from physical immorality. "Her body is a holy temple dedicated by God in which alone may continue the ever complicating warp and woof of evolution," and any gratification of the primitive bestiality of the Negro would cause harm to the race future of the Caucasian.[56]

To deal with the animal passions of the Negro, some doctors prescribed castration. By that method the rapist who prided himself on virility would become an "object of ridicule and contempt" within his own society. After castration the Negro would become "docile, quiet and inoffensive."[57] If executed, he would only be forgotten; castrated and free, he "would be a constant warning and ever-present admonition to others of [his] race." "A few emasculated negroes scattered around through the thickly-settled

[55] The explorer Serres quoted in Paul Broca, *On the Phenomena of Hybridity in the Genus Homo* (London, 1864), 28.

[56] English, "The Negro Problem," 472.

[57] "Castration Instead of Lynching," *Atlanta Journal-Record of Medicine*, VIII (Oct., 1906), 457.

negro communities," argued Dr. Lydston of Chicago, "would really prove the conservation of energy, as far as the repression of sexual crimes is concerned."[58]

As a result of the debilitating effects of syphilis on the Negro, southern physicians felt that his future lay more "in the research laboratory than in the schools." The post–Civil War Negro was a "pitiable creature" whose "mind and body [were] traveling in different directions." He was a product of both development and retrogression, a confusing type. "The negro of 1859 was a fixed type and men could plan with this type as a basis," wrote Dr. Murrell of Richmond. But "the negro of 1889 was a different man—and the negro of today is another."[59] The combination of promiscuousness and its syphilitic progeny, plus the overwhelming responsibility assumed with emancipation, brought the Negro to the brink of insanity. Assuming a "maniacal type," the Negro became a dire threat to civilization in general and southern virtue in particular.[60] While laughter and music were common with the race before the war, emancipation turned the race toward flashy clothes, intemperance, excesses of all kinds, and an accompanying "mental depression and anxiety."[61] Indeed, doctors theorized that the effect of emancipation had been too overwhelming for the race. Having "less mental equipoise" than the white, the Negro suffered "mental alienation from influences and agencies which would not affect a race mentally stronger."[62]

A native of Africa and a savage a few generations ago, then a slave for several generations afterwards; this is the man and the race upon whom the high responsibilities of freedom were thrust; a nation literally born in a day. The history of the world, so far as I know, furnishes no condition similar to that in which the negroes of the South were placed in the first few years after the close of the war. Without education of self or ancestry and without preparation of any sort, the new negro was in-

[58] McGuire and Lydston, "Sexual Crimes," 122–23.
[59] Murrell, "Syphilis and the American Negro," 171.
[60] Bannister and Hektoen, "Race and Insanity," 463.
[61] McGuire and Lydston, "Sexual Crimes," 106; "Negro Traits," *Atlanta Journal-Record of Medicine,* IV (July, 1902), 241.
[62] Miller, "Effects of Emancipation," 290.

vested with the highest functions of citizenship before the healing of the marks of the chains that had bound him.[63]

Few doctors shared the belief that the Negro possessed the capacity for education and civilization in the sense that his physical and intellectual development destined him to the ultimate standard of the American or European. From the very beginning physicians endeavored to set up fixed marks of separation. Some concluded that unless the Negro could "comingle" his blood with whites, his advancement would cease and atrophy at an early stage. This is not to say that they advised miscegenation. On the contrary, they merely used it to show that progress for the Negro race was not possible by any other means. Generally, physicians turned to the Civil War investigations of Gould and Baxter to prove the futility of miscegenation. While miscegenation resulted in increased mental superiority of mixed breeds, it was accompanied by deterioration in physical and moral endowments, leaving them even less adapted than the full-blooded Negro for the struggle of life. The Negro found himself sealed off from progress regardless of the particular scientific explanation adopted.[64]

No matter how one measured the growth of America, it was essentially "the land of the Caucasian," according to most doctors in the late nineteenth century. Those races which entered into the life struggle in America had either to become "caucasianized" or "drop out of the struggle." America, the so-called "botany bay of the world," had no room for races unable to compete in the natural race struggle of civilization.[65]

The implications were ominous for both Negro and white. Only by "disappearing in the mass of population" could the Negro "lose the African cast, and transform himself, by intermarriage and social association, into an actual American."[66] But in the eyes of the white American such a transformation could never be made, since

[63] *Ibid.*, 292.

[64] W. A. Dixon, "The Morbid Proclivities and Retrogressive Tendencies in the Offspring of Mulattoes," American Medical Association, *Journal*, XX (1893), 1–2; Hodges, "Effect of Freedom," 170; Talbot, *Degeneracy, Its Causes, Signs, and Results*, 101–2.

[65] Corson, "Future of Colored Race," 230.

[66] Gilliam, "The African," 440.

the Negro belonged to an alien racial stock. Doctors generally disapproved of forced migration out of the country. "Leaving ethical considerations out of the question," wrote Dr. Charles S. Bacon of Chicago, "3,000,000 workers form too valuable an economic factor to be eliminated unless the race problem is too dangerous to the state and there is no possibility for solving it in any other way."[67] While utilizing his manual labor, they had to eliminate the Negro as a political and social threat.[68]

> The ballot will not make a man moral, industrious, thrifty, or healthy—the basic qualities of success. It is true that bad laws, for instance unjust taxation, may interfere with a man's or a nation's progress. But to the healthy, industrious man these obstacles will not be insuperable, and moreover the ballot in the hands of the ignorant man will not by some magical process bring good laws. . . . I believe the best advisers of the [Negro] race counsel its members to pay attention to their own business, study their condition and opportunity, and leave the saving of the country to others.[69]

It would be dangerous to bring Negroes into the political process, argued sociologist Gilliam, not simply because they were inferior but also because, being distinct, they would "stand together socially" and their distinctions would "morally compel them to stand together politically." Confined by a social barrier, the Negro would "develop abnormally the natural race-instinct, and under a powerful esprit de corps, cast a solid ballot."[70]

Education was no solution to the Negro's race inferiority, argued Dr. Bacon. "A classical education for a negro whose proper vocation is raising rice or cotton or garden truck, is as much out of place as a piano in a Hottentot's tent."[71] The Negro had to be treated wholly as a "parasite," added Dr. E. T. Brady of Abingdon, Virginia. Some Negroes had been educated in the North, leading liberal thinkers to believe them capable of making ethical and moral judgments. But this was not true. "They are just as devoid of

[67] Charles S. Bacon, "The Race Problem," *Medicine* (Detroit), IX (May, 1903), 342.

[68] Gilliam, "The African," 441.

[69] Bacon, "The Race Problem," 342.

[70] Gilliam, "The African," 441.

[71] Bacon, "The Race Problem," 342.

ethical sentiment or consciousness as the fly and the maggot."[72] Negroes had to remain in the South, give up their "aspirations to full citizenship," and hand their education and government over to the Caucasian. They must remain the "hewers of wood and drawers of water." Furthermore, added Dr. Hodges of the College of Medicine in Richmond, "some kind of restraining and inhibitory influences, such as once characterized the institution of slavery, must be thrown around [them] as a safeguard for many years to come."[73]

FREDERICK L. HOFFMAN

In 1896 the American Economic Association published "Race Traits and Tendencies of the American Negro," a study by Frederick L. Hoffman (1865–1946), a statistician who worked for the Prudential Insurance Company of America. Hoffman had published earlier studies concerning the Negro in *Arena* (1892), *Medical News* (1894), and the *Publications* of the American Statistical Association (1895). A member of the American Academy of Medicine, the American Statistical Association, and the Royal Statistical Society of London, Hoffman, in his work on the racial characteristics of the American Negro, reflected a summation of the century's medical and anthropological accumulations concerning racial relations in America. Hoffman's conclusions mirrored the cumulative tendencies of a century of American and European medical and somatometric studies on race.[74]

Hoffman took issue with those census alarmists like E. W. Gilliam, whose figures indicated that the Negro population in the United States during the 1880's was increasing at a faster rate than the Caucasian. From 1800 to 1890, Hoffman argued, the percentage of increase for the white population rose from 81.12 to 87.80 percent while the Negro in the same period declined from

[72] Quoted in Murrell, "Syphilis and the American Negro," 173.

[73] Hodges, "Effect of Freedom," 171.

[74] The American Economic Association organized in 1885. Its officers included Francis Amasa Walker, an outspoken critic of immigration and superintendent of the Ninth and Tenth Censuses, Charles F. Dunbar, John B. Clark, Franklin H. Giddings, Jeremiah W. Jenks, F. W. Taussig, and Davis R. Dewey.

18.88 to 11.93 percent.[75] He pointed out that Gilliam's statistics considered only birth-rate figures and took no cognizance of the death rate of the particular race stock. Though Hoffman admitted that the birth rate among Negroes was in excess of Caucasian natality, nonetheless, Negro mortality far exceeded its own birth rate.[76] Rhode Island, Connecticut, and Massachusetts all reported an excess of Negro deaths over natality, and similar statistics were available for such cities as Washington, Baltimore, Richmond, Memphis, Louisville, Atlanta, Savannah, Charleston, Mobile, and New Orleans.[77] What the statistical evidence implied was that while the Negro race had a higher birth rate, its movement from the plantation to the city, as well as its change from slavery to freedom, had undermined both its health and its race future. Outside the artificial framework of a slave system, which preserved the Negro stock in a "hothouse" condition, the race could neither maintain itself nor perpetuate its meager achievements.[78]

It became apparent, as Hoffman's inquiry developed through more than 300 pages of statistics and syntheses of pre-Darwinian, evolutionist, and medical investigations, that the efforts of the higher races to ameliorate the condition of the Negro or, for that matter, any of the lower races, had the effect of exaggerating the differences between the races. After thirty years of freedom, Negro and Caucasian were "farther apart than ever in their political and social relations."[79] In order to determine the actual degree of dif-

[75] Frederick L. Hoffman, "Race Traits and Tendencies of the American Negro," American Economic Association, *Publications,* XI (Aug., 1896), 6.

[76] *Ibid.,* 33–34; T. N. Chase, "Mortality among Negroes in Cities," Atlanta University, *Publications,* I (1903), 1–24; Walton, "Comparative Mortality," 291–94.

[77] Hoffman, "Race Traits and Tendencies," 35, 39.

[78] George S. Painter, "The Future of the American Negro," *American Anthropologist,* n.s., XXI (Oct.-Dec., 1919), 410. In May, 1896, Atlanta University, reflecting the Negro concern over mortality statistics, held a conference on the subject of "Negro Mortality in Cities." See U.S. Department of Labor, "Condition of the Negro," 257; J. Bradford Laws, "The Negroes on Cinclare Central Factory and Calumet Plantation, Louisiana," U.S. Department of Labor, *Bulletin,* no. 38 (1902), 119.

[79] Hoffman, "Race Traits and Tendencies," 1. See Albert B. Hart, *The Southern South* (New York, 1910), 10–11. Hart called Hoffman's statistical findings "very widely read and quoted in the South." See also Nathaniel S. Shaler, "Science and the African Problem," *Atlantic Monthly,* LXVI (July, 1890), 37, 42.

ference between the white and black races, Hoffman turned to the evidence of race vitality to learn whether the Negro had undergone change from the time of his servitude to that of a freedman. His effort to ascertain such a change developed out of the seemingly "indisputable evidence" of physicians and statisticians in the 1880's and 1890's that the Negro showed "the least power of resistance in the struggle for life."[80] Though he argued that the Negro race had an excessive mortality rate, he discounted all arguments that placed blame or causation on the low social and economic conditions of the people. It was impossible to accept the argument that "given the same social, economic and sanitary conditions of life, the colored race would enjoy the same health and favorable death rate as the white population."[81] He drew upon the evidence of Dr. John Moore, Surgeon-General of the United States Army, and Dr. R. M. Cunningham of the Alabama penitentiary, who argued that "even under the same conditions . . . the negro is still subject to a higher death rate."[82]

Hoffman, agreeing with physicians, expressed his belief that the Negro prior to the Civil War "enjoyed health equal if not superior to that of the white race."[83] Borrowing his terminology from Englishman Benjamin Kidd, Hoffman contended that the new generation of Negroes in America showed the greatest loss of "social effectiveness."[84] To substantiate his personal findings, Hoffman delved into the Civil War Sanitary Commission anthropological investigations of Gould, the somatometric reports of the Provost Marshal–General's Bureau, and the study of Dr. Sanford B. Hunt on Negro soldiers. There was abundant proof in their investigations, Hoffman argued, that the post–Civil War generations of Negroes were more liable to disease than their prewar ancestors.

80 Hoffman, "Race Traits and Tendencies," 37; Tiffany, "Comparison between Surgical Diseases," 261–73; Powell, "Increase of Insanity," 1185–88; F. H. Fetter, "Race Degeneration and Social Progress," *Forum*, XXVIII (Oct., 1889), 228; Edward B. Reuter, *The Mulatto in the United States* (Boston, 1918).

81 Hoffman, "Race Traits and Tendencies," 49.

82 *Ibid.*, 50.

83 *Ibid.*, 55.

84 *Ibid.*, 57; Monroe N. Work, "Crime among the Negroes of Chicago," *American Journal of Sociology*, VI (Sept., 1900), 222; Bruce, *Plantation Negro as a Freeman*, 243.

For Hoffman, it meant that the Negro soldier, forced for the first time to undergo as an equal the hardships and problems of the white soldier, "showed a higher mortality rate while subjected to the same, or perhaps more favorable conditions."[85]

Again Hoffman looked to Civil War anthropometric statistics to demonstrate the "lower vital power" among Negroes. Comparing the war findings with later medical and anthropological statistics on the same bodily parts, he conjectured that the discrepancy gave conclusive evidence of a degeneration in the Negro population.[86] Subsequent anthropometry by the medical departments of the New York Life Insurance Company in 1874 and 1895, the Washington Life Insurance Company in 1886, the Prudential Insurance Company of America in 1895, and the statistics drawn up in 1893, 1894, and 1895 by the United States Army showed a decline in physiological capacity since the Civil War findings.[87] Decrease in chest expansion, decrease in the size of the thorax, increase in consumption and respiratory diseases, smaller weight of the Negro lung, "mean frequency of respiration" which was greater in the Negro than in the white, and inferior power of vision in the Negro "prove conclusively that there are important differences in the bodily structure of the two races, differences of far-reaching influence on the duration of life and the social and economic efficiency of the colored man."[88] Agreeing with Sir Duncan Gibb of the London Anthropological Society that "the vital energies of a people had a great deal to do with the state of the body, and that the capacity of the chest should count for something very considerable as an indication of national power," Hoffman predicted a fateful end to Negro aspirations.[89]

> The general conclusion is that the negro is subject to a higher mortality at all ages, but especially so at the early age periods. This is largely the result of an inordinate mortality from constitutional and respiratory diseases. Moreover, the mortality from these diseases is on the increase

[85] Hoffman, "Race Traits and Tendencies," 99; Hunt, "The Negro as a Soldier" (1869), 40–54.
[86] Hoffman, "Race Traits and Tendencies," 162.
[87] *Ibid.*, 149; Laws, "The Negroes," 119–20.
[88] Hoffman, "Race Traits and Tendencies," 171.
[89] Hoffman quoting Gibb, *ibid.*, 171–72.

among the colored, and on the decrease among the whites. In consequence, the natural increase in the colored population will be less from decade to decade and in the end a decrease must take place. It is sufficient to know that in the struggle for race supremacy the black race is not holding its own; and this fact once recognized, all danger from a possible numerical supremacy of the race vanishes. Its extreme liability to consumption alone would suffice to seal its fate as a race.[90]

The lack of suicides among the Negro race was cited as another instance of inferior physiological and psychological organization. Only in rare cases did the Negro commit suicide, and even then it was generally "only in a fit of passion, during loss of self control, or as in most cases, to escape the consequences of his crimes."[91] In no instance could Hoffman see traces of those "more subtle motives" which prompted "the . . . more cultured and more advanced races."[92] Much of the reasoning behind the lack of suicide grew from the absence of anxiety in the Negro organization and "his tendency to live wholly in the present." The incompleteness of his intellectual development deprived him of the "coolness and fortitude" that was lacking in the inferior races, despite the degree to which he mimicked superior civilizations.[93]

Hoffman believed that intermarriage among races of similar culture resulted in physical and psychical advantages for both stocks, but that mixtures of Germans and Italians, English and Spaniards, Swedes and Turks, let alone Caucasians and Negroes, was an altogether different matter.[94] Concerned primarily with the crossing of white and black, Hoffman emphasized that the product was inferior both physically and morally to the organization of both parents. Agreeing with earlier conclusions of polygenist Josiah Nott, he argued that the mulatto was "possessed of the least

[90] *Ibid.*, 148; John B. Haycraft, *Darwinism and Race Progress* (London, 1900), 52–55.

[91] Hoffman, "Race Traits and Tendencies," 140–41.

[92] *Ibid.*, 142.

[93] Hoffman quoting Bruce, *Plantation Negro as a Freeman*, in *ibid.*, 143.

[94] *Ibid.*, 179–80; Ray S. Baker, "The Tragedy of the Mulatto," *American Magazine*, LXV (1908), 582–98; Alfred H. Stone, "The Mulatto Factor in the Race Problem," *Atlantic Monthly*, LXI (May, 1903), 658–62; Richard Mayo-Smith, "Theories of Races and Nationalities," *Yale Review*, III (1894), 166–85.

vital force" of all races.[95] To substantiate Nott's findings, Hoffman went to the opinions of northern physicians in J. H. Baxter's *Statistics, Medical and Anthropological,* who with near unanimity agreed that the mulatto was least capable in army life and most susceptible to physical disability. Despite the mulatto's undoubtedly superior intellectual capacity over the pure black, a situation verified from Sanford Hunt's investigations of brain weight after autopsy, the increasing intellectuality in no way compensated for the overburdening deterioration in physical and moral capacity. On the strength of such experiments, Hoffman concluded that miscegenation was detrimental to the true progress of both white and black and resulted in an "inferior social efficiency and diminishing power as a force in American national life."[96]

> Hence the conclusion is unavoidable that the amalgamation of the two races through the channels of prostitution or concubinage, as well through the intermarrying of the lower types of both races, is contrary to the interest of the colored race, a positive hindrance to its social, mental and moral development. But aside from these considerations, important as they are, the physiological consequences alone demand race purity and a stern reprobation of any infusion of white blood. Whatever the race may have gained in an intellectual way, which is a matter of speculation, it has been losing its greatest resources in the struggle for life, a sound physical organism and power of rapid reproduction.[97]

Hoffman sought particularly to deprecate the philanthropic and educational attitudes of the late nineteenth century. Any benefit the Negro received by way of white educational processes influenced in no way the moral progress of the race. The Negro race, despite individual examples to the contrary, "has gone backwards rather than forwards." White humanitarianism had deprived the Negro of the merits and virtues inherent in self-help.[98] Any effort

[95] Hoffman, "Race Traits and Tendencies," 182; Jabez L. M. Curry, "The Negro Question," *Popular Science Monthly,* LV (June, 1899), 178. Curry was general agent of the Peabody Education Fund and of the John F. Slater Education Fund.

[96] Hoffman, "Race Traits and Tendencies," 187–88; Van Evrie, *White Supremacy and Negro Subordination,* 148.

[97] Hoffman, "Race Traits and Tendencies," 206–7.

[98] *Ibid.,* 236.

to improve the condition of the lower races without "the vital element of self-help" would prove a failure. "A system of philanthropy," he wrote, "that is based on the notion that easy conditions of life are essential to human development must fail in its effort, honorable and unselfish as the motives may be."[99] In effect, Hoffman and others like him in the late nineteenth century were criticizing the methods of philanthropy. Essentially, they argued that the Caucasian, superior in organization, had far surpassed the inferior races in moral attitudes, benevolence, and humanitarianism. Yet the basis of his ethical and highly humanitarian efforts to raise the inferior races to his own standards had, in effect, destroyed the capacity for self-improvement in the inferior races. The Caucasian, who developed from his own self-help "under the most adverse circumstances," deprived those beneath him of the same essential evolutionary ingredients. The Caucasian had really been the only man in evolution since, from the very beginning of his superior development, he sought to reach out and help those beneath him—a situation which unconsciously and inadvertently had harmed other race stocks.[100]

Hoffman suggested what amounted to a separation of the races or a laissez-faire approach to race relations. It was useless, he argued, to make the Negroes into anything which they were unable to achieve or maintain without the Caucasian's help. Left to themselves, "the great majority leave the earth as poor as they entered it, and are fully satisfied with a degree of comfort too low to prove of economic advantage to the state."[101] Advance for the inferior races of man could only become permanent as a result of virtues achieved by the races themselves through their own effort and struggle in life. Thirty years of freedom in the United States and nearly sixty years in the West Indies had failed to extinguish those racial differences which the abolitionist had explained away in the privation of freedom. Despite freedom, the moral, mental, and economic level of the Negro race remained far below the

[99] *Ibid.*, 141–42.
[100] *Ibid.*
[101] *Ibid.*, 308.

superior races.[102] Sensing that the evil lay in philanthropy, Hoffman called for a halt to the "modern attempts of superior races to lift inferior races to their own elevated position," since the result was almost criminal in its interference with the natural order of race struggle among nations and peoples.[103]

While the Aryan confronted directly the conditions of life in a struggle for existence and transmitted his qualities to succeeding generations, the inferior races, eliminated from the ranks of struggle by overzealous humanitarian efforts, became sterile contributors to their race future. The "easy conditions of life," wrote Hoffman, added to "a liberal construction of the doctrine of forgiveness of sins and an unwarranted extension of the principle of state or private interference in the conduct of individual life," never had nor ever would raise a race of inferior people to a higher plane.[104] Modern educational and philanthropic enthusiasm, he judged, had succeeded in making the Negro race more dependent upon the Caucasian than in the days prior to emancipation. This downward-spiraling tendency of the Negro race could be arrested only by radical changes in race relationships. If not, the time would come, he predicted, when diminished vitality, morality, and economic efficiency would bring about the Negro's extinction. First and foremost in any solution was the necessity of bringing an end to the white man's aid and assistance. The Negro, if he was to be anything other than an artificial creation of the white man's philanthropy, must refuse and be refused "every offer of direct interference in his own evolution."[105]

As late as 1910 Hoffman still accepted the substance of his earlier investigations. In a book written by Edward Eggleston, Hoffman was quoted as still believing that the Negro race was of a basically inferior constitution. Medical science had removed the

[102] *Ibid.*, 311; Frederick L. Hoffman, "The Negro in the West Indies," American Statistical Association, *Publications*, IV (1895), 199–200.

[103] Hoffman, "Race Traits and Tendencies," 312; B. S. Coler, "Reform of Public Charity," *Popular Science Monthly*, LV (Oct., 1899), 750–55.

[104] Hoffman, "Race Traits and Tendencies," 327.

[105] *Ibid.*, 328; Robert H. Lowie, "Educational Theories," *American Anthropologist*, n.s., XIV (Apr.-June, 1912), 395–99.

possibility of the race's ultimate extinction, yet his correspondence with southern physicians as well as his own investigations convinced Hoffman that, at most, the Negro race would become like the gypsy of Europe, an anachronism of modern civilization, existing on the fringe of society and neither contributing to nor detracting from civilization's progressive development.[106]

Physicians were generally agreed on the condition of the Negro in the late nineteenth century. Arguments to the contrary were simply not to be found in the transactions and journals of the medical societies. Expressing the quiet intimacy of a consulting-room conversation, doctors exhausted all possible arguments in their commentaries on the Negro and his health. They vehemently dismissed the possibility for race improvement and, with a minimum expenditure of rhetoric, they offered a prophetic warning for the race's future.

[106] Hoffman to Eggleston, Aug. 5, 1910, in Edward Eggleston, *The Ultimate Solution of the American Negro Problem* (Boston, 1913), 272–73.

III *The Species Problem: The Origin of Man Controversy*

While most anthropologists in the nineteenth century busily engaged in the technical aspects of somatometry, a good number of them were also concerned with the more problematic question of man's origin. Like somatometry, the speculation into origination grew out of the awareness of differences in the broad spectrum of genus *Homo.* The taxonomic system of Linnaeus precipitated not only an intensive study of comparative structures but it also led to the question of whether the various "races" of man had origin in one primitive stock. Were Negroes, Hottentots, Eskimos, and Australians really men in the full sense of the term, sharing in the intellectual endowments of Europeans, or were they half brutes, not belonging to what French scientist Bory de Saint-Vincent called the "Race Adamique"? Defined in other terms, the problem concerned whether humanity descended from a single monogenistic type, or whether humanity had distinct polygenistic ancestors. If it were true that these peoples were really half brutes, then, some argued, they should become subject to the superior races. The subsequent controversy between monogenists and polygenists became the longest of the internecine battles among the scientists of man.

THE MONOGENISTS

The monogenists, divided even among themselves, by no means propounded a single theory. One group, the Adamites, held strictly to the biblical epic of creation. They accepted literally the story of Adam and Eve and explained the races of the world as having descended from the eight people who survived the Deluge and landed on Mount Ararat. Science in no way entered into the origination of man for these Adamite adherents. Theirs was really not a theory; rather, it was an article of faith.

A second faction among the monogenists tried to accommodate themselves to both the biblical Adamites and the developments of science. They produced a hybrid interpretation, a combination of liberal Christianity and the higher criticisms of science. Sometimes called the rational monogenists, they included within their ranks such men as Linnaeus, Georges Buffon, Georges Cuvier, Blumenbach, James Cowles Prichard, and Armand de Quatrefages. Generally they held that the earth was much older than the biblical epic, that man had been created somewhere between the Caucasus and the Hindu Kush, and that the differences in man were due to "the existing diversities of climate and other conditions" that acted upon the waves of migration leaving this original homeland. Recognizing but one human species, the rational monogenists saw the human races as varieties arising from the influence of such environmental factors as climate, although they did not wholly discount the possible intervening influence of the Supreme Will.[1]

[1] Thomas H. Huxley, *Man's Place in Nature and Other Anthropological Essays* (New York, 1894), 142; C. Staniland Wake, "The Adamites," Anthropological Institute of Great Britain and Ireland, *Journal*, I (1871–1872), 363–76; Thomas Bendyshe, ed., *The Anthropological Treatises of Johann Friedrich Blumenbach* (London, 1865), 264–65; M. Flourens, "History of the Works of Cuvier," Smithsonian Institution, *Annual Report for 1868*, 141–65; Gustaf Retzius, "The So-Called

The differences in ideas of race inferiority between the Adamites and the rational monogenists were hardly as grave as their differences over the Caucasian prototype. The Negro in the biblical explanation of race differences was the result of the curse of Ham, while for the rational monogenist, color and inferior physiological development stemmed from a scientific belief in degeneracy.[2] Despite a difference in methodology, their conclusions were strikingly similar. Both theories, one representing the religious establishment and the other a liberal-Christian consensus, appeared as contrived rationalizations of *a priori* judgments.

Another school of monogenists, the transformists, were of the French Normal School, and derived the substance of their ideas from the theories of Jean Lamarck. For the transformists, species, "considered as regards time," did not exist. Developing from "a small number of primordial germs or monads, the offspring of spontaneous generation," species passed through successive transformations or divergences. Men, the offspring of a slow transformation of apes, were "isolated extremities of the branches and boughs" of the organic kingdom.[3] Cuvier and the more orthodox of the monogenists ridiculed the ideas of the transformists, and as a result their views were sorely abused in the years before Darwin. Yet, despite Cuvier's opposition, adherents included such men as Bory de Saint-Vincent, Lorenz Oken, Herbert Spencer, and Charles Lyell.

It was long held among the monogenists that the races sprang from a single family and that differences in color, body form, and intelligence were the result of environmental changes affecting the migrant stock as it adapted itself over many generations. Monogenist theorists did not recognize the existence of "pure" races but only the relative permanence of marked varieties suited to different regions and gradually produced by the inheritance of acquired

---

North European Race of Mankind," Anthropological Institute of Great Britain and Ireland, *Journal*, XXXIX (1909), 277–313; Thomas Bendyshe, "The History of Anthropology," Anthropological Society of London, *Memoirs*, I (1863–1864), 335–458.

[2] Alexander Winchell, *Preadamites: Or, a Demonstration of the Existence of Men before Adam* (Chicago, 1888), v, 271–72; Bendyshe, ed., *Anthropological Treatises*, x–xi.

[3] Paul Topinard, *Anthropology* (London, 1878), 519–20.

variations through the influence of external, environmental conditions, "fixed" (but not absolutely) through centuries of close breeding. The permanent varieties of man found in the world, argued monogenist Prichard, differed from species "in that the peculiarities [were] not coeval with the tribe, but [arose] since the commencement of its existence."[4]

Monogenists defined the difference between race and species by means of the terms "hybrid" and "mongrel." If the Negro and the Caucasian, for example, represented two varieties (or races) of a single species, then the result of their union would be a "mongrel" whose generative faculties would be equal to that of the parents. If, however, Negro and Caucasian represented two distinct species, their union would result in a sterile "hybrid." What monogenists sought to determine was whether mongrels or hybrids were the products of human mixtures. If mongrels were the product, then their mixtures were a common occurrence and their fertility in the first generations was equal to that of their parents.[5] If hybrids were the product, then the generative faculties would be greatly reduced. "If two of these first hybrids are united they produce hybrids of the second generation," wrote monogenist Quatrefages. "In most cases, however, the latter are either sterile, or present the phenomenon of a spontaneous return to one or the other of the parent type."[6] For Quatrefages, the crossing of hybrids did not produce a race but "only . . . varieties incapable of transmitting their individual characters."[7] Hybrids were not products of natural forces. Left to themselves, two separate species never mixed.

Monogenists argued against the belief that the crossing of different races would bring sterility or infertility. On the contrary, "fertility is the law of union between animals belonging to different

[4] Charles Hamilton Smith, *Natural History of the Human Species* (Boston, 1851), 22.

[5] Armand de Quatrefages, *The Human Species* (New York, 1879), 70–71.

[6] *Ibid.*, 72–73; G. O. Groom Napier, "Notes on Mulattoes and Negroes," Anthropological Society of London, *Journal*, VI (1868), lvii–lx.

[7] Quatrefages, *The Human Species*, 74; Earl W. Count, "The Evolution of the Race Idea in Modern Western Culture during the Period of the Pre-Darwinian Nineteenth Century," New York Academy of Science, *Transactions*, VIII (1946), 139–65.

races."[8] The European had crossed with practically every known race in his conquest of the world. These unions resulted, in "certain parts of the globe, and notably in America, [in] an inextricable mass of mixed peoples," wrote Quatrefages, "perfectly comparable with our street-dogs and roof-cats."[9] Many monogenist arguments for unity of species grew out of the experiments and observations of John Bachman, minister of St. John's Lutheran Church in Charleston, South Carolina.[10]

Despite their insistence on man's single origin, the monogenists were not egalitarians. Races, during centuries of formation, acquired characteristics that, upon comparison, established an inequality "impossible to deny." The Negro had never been equal to the white. Quatrefages wrote, "Does it follow that, because all the races of dogs belong to one and the same species, they all have the same aptitudes? Will a hunter choose indifferently a setter, or a bloodhound to use as a pointer or in the chase? Will he consider the street-cur as of equal value with either of these pure-breeds? Certainly not. Now we must never forget that, while superior to animals and different to them in many respects, man is equally subject to all the general laws of animal nature."[11]

Though the "radical" Jeffersonians in American society were inclined to favor monogenism as providing the best scientific or religious certainty for the "self-evident" truth "that all men are created equal," this was hardly more than an inclination on the part of the monogenist himself. Monogenists had drunk deep of the effects of environmentalism and saw no reason to conclude that the Negro was anything but inferior. In fact, rather than to deny inequality, one could argue that the theory of monogenism grew out of an *a priori* belief in degradation from the original prototype.

[8] Armand de Quatrefages, *The Natural History of Man* (New York, 1875), 78.

[9] *Ibid.*, 29.

[10] John Bachman, "An Investigation of the Causes of Hybridity in Animals on Record, Considered in Reference to the Unity of the Human Species," *Charleston Medical Journal*, V (1850), 168–97; Bachman, *The Doctrine of the Unity of the Human Race Examined on the Principles of Science* (Charleston, S.C., 1850), 80–81; William Stanton, *The Leopard's Spots: Scientific Attitudes toward Race in America, 1815–59* (Chicago, 1960), 123–36.

[11] Quatrefages, *The Human Species*, 450–51.

As an environmentalist, the monogenist had drawn his schema of race classification from changes in the genus *Homo.* His race classification and derivation of race stocks, drawn from his measuring instruments, contained built-in feelings of superiority. True, the Negro might have been born equal, but the monogenist of pre-Darwinian years had no desire to carry the statement as far as the "radical" Jeffersonian would have liked. Man's being born equal, the monogenist argued, had little consequence if the various "races" did not remain equal. Except that all men were born equally men, the axiom of Jefferson was meaningless rhetoric.

THE POLYGENISTS

Like the monogenist school the polygenists were neither monolithic nor overly consistent in their theoretical stance. One segment, the neotraditionalist school, adhered to the biblical story; yet at the same time it tried to account for the various types of mankind. Though decidedly Christian in its orientation, this neotraditionalist school felt it necessary to reconcile Scripture with polygenism. There were other peoples, argued polygenist Paul Broca, who existed along with the Adamite family, "with whom the sacred writer had no concern."[12] Adam and Eve in this new formulation referred only to the Jewish race. This school held that men and animals were created essentially where they were found, which meant multiple creation. In other words, man emerged in several places by several special acts of creation, and the various forms were distinct. Neotraditionalists like Louis Agassiz, Lord Henry H. Kames, and Karl Vogt prepared the groundwork for Darwin by showing modification of types through creative changes, but they found no indication from paleontology of evolution from a single protoplast or change within geological periods.[13]

[12] Paul Broca, *On the Phenomena of Hybridity in the Genus Homo* (London, 1864), 67.

[13] Karl C. Vogt, *Lectures on Man, His Place in Creation, and in the History of the Earth* (London, 1864), 464–65; Henry H. Kames, *Six Sketches of the History of Man* (Philadelphia, 1776); Louis Agassiz, "Professor Agassiz on the Origin of Species," *American Journal of Science and Arts,* XXX (1860), 148; Jules Marcou,

A second polygenist division accepted the conclusions of the neotraditionalists but adhered more strongly to the strictures of Mosaic cosmology. They felt that the lapse of biblical time, which they ascertained to be 5,877 years, was insufficient to produce the conditions of race and, hence, man's origins had been multiple. Accepting the Mosaic cosmology, polygenism could explain the varieties of man only by separate and special creations, as the time span was far too short to permit the necessary changes in human varieties to occur, either from degeneration or dispersion.

A third polygenist school was related to the Lamarckian theorists of monogenism. These adherents held that the various races of men resulted from modification "of some antecedent species of ape—the American from the broad-nosed Simians of the New World, the African from the Troglodyte stock, the Mongolian from the Orangs."[14] These Lamarckian polygenists saw the geological barriers as so formidable that they prevented migration from a single center. Contrary to the Lamarckian monogenists, they found it far easier to derive the American Indian, African, and European within regional limits and from different species of apes.[15]

Early theorists of the polygenist schools favored the term "species" in their belief in the diversity of man. In the context of their definition species were "fixed" and did not naturally cross with other species, except under artificial conditions. Although there was occasional fertility between the separate species, the product of the union was sterile or tended toward sterility, proving the "unnaturalness" of the original union. The concept of species was important to those scientists in the nineteenth century who drew their schematization of the universe from the logical and spatial arrangement of the Chain of Being. For if one hybrid were capable of increase, the divine arrangement of the Creator would have

---

Life, Letters, and Works of Louis Agassiz, 2 vols. (New York, 1895), II, 125; Edward Lurie, *Louis Agassiz: A Life in Science* (Chicago, 1960), 161–63; Lurie, "Louis Agassiz and the Races of Man," *Isis*, XLV (1954), 227–42; Juan Comas, *Manual de anthropología física* (Mexico City, 1957), 105.

14 Huxley, *Man's Place in Nature*, 142.

15 John L. Myres, "The Influence of Anthropology on the Course of Political Science," University of California, *Publications in History*, IV (1916), 71.

been distorted and a destructive imbalance set into the order of the world. All living things formed one chain of universal being from the lowest to the highest. None of the species originally formed were extinct. Nature proceeded according to divine plan and admitted of no improvement. The continuation of this belief into the nineteenth century precipitated an enormous amount of speculation on whether the mulatto was more or less fertile than either of the two original stocks. The general consensus was that the mulatto was less fertile and, hence, an artificial "hybrid" tending toward extinction. As the polygenists of the nineteenth century turned to the term "race" rather than "species" as the definition of human types, so they borrowed the word "mongrel" in exchange for "hybrid" to identify the offspring of mixing. In doing so, however, they created a confusion in terminology since the monogenists' criteria for "species," "race," "mongrel," and "hybrid" remained unchanged.[16]

In the decades before the American Civil War polygenists obtained their most vocal supporters from the United States. Adherents like Dr. Charles Caldwell, Dr. Samuel George Morton, George R. Gliddon, Governor James Henry Hammond of South Carolina, Josiah C. Nott, Louis Agassiz, Peter A. Browne, and the sympathetic support of William Gilmore Simms's *Southern Quarterly Review* and *De Bow's Review* gave both political and scientific weight to the theory. For these men, both the American Indian and the Negro were true autochthons of their respective continents. There was no link between the Old and New Worlds, and any appearance of similarity was far outweighed by the multitude of physical, moral, and mental differences.[17] Agassiz, for example,

[16] Vogt, *Lectures on Man*, 441; Thomas H. Huxley, "What Are Species?" *Popular Science Monthly*, IX (Aug., 1876), 412.

[17] Irving A. Hallowell, "The Beginnings of Anthropology in America," in Frederica de Laguna, ed., *Selected Papers from the American Anthropologist, 1888–1920* (New York, 1960), 65–66; T. D. Steward and M. T. Newman, "An Historical Resume of the Concept of Differences in Indian Types," *American Anthropologist*, n.s., LIII (1951), 28; Samuel G. Morton, *An Enquiry into the Distinctive Characteristics of the Aboriginal Nations of North and South America* (Philadelphia, 1842), 6; Morton, *Brief Remarks on the Diversities of the Human Species* (Philadelphia, 1842), 21; Morton, *Crania Aegyptiaca: Or Observations on Egyptian Ethnography, Derived from Anatomy, History and the Monuments* (Philadelphia,

believed that the divisions of mankind were primarily distinct and not originating from one primordial form. The branches of mankind as well as of the animal kingdom were "founded upon different plans of structure, and for that very reason have embraced from the beginning representatives between which there could be no community of origin."[18] The offspring of Caucasian and Negro were "hybrids" and characterized by either sterility or reduced fecundity.[19]

The school of polygenism, however, did not hold a monopoly on race inferiority or the proslavery argument. In America as in Europe concepts of race inferiority existed in both monogenist and polygenist schools. The polygenists gained temporary notoriety in the pre–Civil War years because of their insistence that the Negro was not only a separate species but was incapable of modification through time. Environmental change, they argued, offered an optimistic palliative but took no cognizance of the fact that the Negro had remained unchanged through centuries of breeding. Not only his inferior physiological characteristics but also his social status as a slave remained unchanged from the time of the Egyptians to the days of slavery in the South.[20] Inferiority was a permanent stain on the race and marked the Negro for slave status. But the monogenists, despite their insistence on environmental change through time, were no more favorable to the Negro, except in their remote theoretical stance. For all practical purposes, monogenists accepted the known race stocks as "fixed" as a result of centuries of inbreeding. Change in the Negro's status, if it were to take place, would require undetermined generations and influences. Almost the whole of scientific thought in both America and Europe in the decades before Darwin accepted race inferiority, irrespective of whether the races sprang from a single original pair or were created separately. Whether for or against slavery, anthropologists

---

1844), 66; James H. Hammond, *Selections from the Letters and Speeches of the Hon. James H. Hammond, of South Carolina* (New York, 1866), 114–20.

18 Agassiz, "On the Origin of Species," 143.

19 Elizabeth Agassiz, ed., *Louis Agassiz: His Life and Correspondence* (Boston, 1885), 598.

20 Morton, *Crania Aegyptiaca*, 66; Josiah C. Nott, *Instincts of Races* (New Orleans, 1866), 102–18.

could not escape the inference of race subordination, either in the monogenist degeneracy theory of Blumenbach or the polygenist stance of Louis Agassiz.

With the exception of Louis Agassiz, those polygenists who voiced their opinions the loudest in America were also the more notorious for their anti-biblical language. The works of Josiah Nott and George Gliddon were as much involved in "parson-digging" as with the origins controversy. By the same token, however, it was just such "parson-digging" that limited the scope of their appeal to a very narrow sector of American society. On the whole, the scientific community in America harbored little disaffection with the symbols of religion. Then, too, the arguments of the polygenists, expressed by the school's most vocal adherents, were far too secular and confusing for a generation moving slowly toward military confrontation. The South was too fundamentalist and New England too moralistic to meet on scientific terms which were unbiblical and unemotional. The aggressive character of northern abolitionism and southern expansionism forced the two sectional combatants into a dialectical position of the lowest common denominator—a position that was common to both sides and capable of the largest emotional appeal. The stance of both North and South was basically Christian, biblical, and monogenist. The scientific argument of diverse origin, by reason of its generally more anti-biblical approach, moved more and more out of the public eye and back into the closed circle of a few scientific savants. "It is not from the writings of polygenists," wrote Paul Broca, "but from the Bible, that the representatives of the Slave States have drawn their arguments." The monogenist theory, interspersed with biblical and scientific description, became the common funding source of American sectional differences.[21]

It was the Englishman Thomas Huxley's contention that Darwin's *Origin of Species* (1859) brought the age-old controversy between monogenists and polygenists to a close. Darwin's hypothesis, he argued, gathered together monogenists and polyge-

[21] Broca, *Hybridity in the Genus Homo*, 70; Broca, "The Progress of Anthropology in Europe and America," Anthropological Institute of New York, *Journal*, I (1871–1872), 22–42.

nists on a far different plane of understanding. Darwin had shown, according to Huxley's interpretation, that the premises of both schools could be accepted without necessitating an acceptance of their respective conclusions. "Admit that Negroes and Australians, Negritos and Mongols are distinct species, or distinct genera, if you will," he wrote, "and you may yet, with perfect consistency, be the strictest of monogenists, and even believe in Adam and Eve as the primaeval parents of all mankind." Huxley was anxious that the anthropologists move on in their study of man. Far too much time had been wasted on the origination controversy when valuable research needed to be done in the classification of races through cranial, hair, and skin measurements.[22]

But the origins controversy took a different course in scientific discussions in America after 1860. It was a course that resulted as much from the impact of the Civil War as from the publication of Darwin's *Origin of Species.* The political climate around the Negro in America during and after the Civil War—the "favored race," as scientist Joseph LeConte called it somewhat cynically—brought the origins controversy to a temporary halt among American naturalists. A political and military solution, implemented by the Thirteenth, Fourteenth, and Fifteenth Amendments, a civil rights act, and several force bills, had answered the Negro question and had established through law his position in the order of American society. It was the final blow to those polygenist scientists who had lent their names to the politics of the prewar era. The American school of polygenism was scathingly rebuked and accused of scientific casuistry in making the Negro a separate species to soothe a southern rationale.

JOSIAH C. NOTT

Alabama scientist Josiah Nott, the last of the vocal polygenists of the prewar period, reluctantly accepted Darwinism. His adherence at first seemed due much more to the dysteleology of Darwin's program of natural selection, for as he wrote to Ephraim Squier in

[22] Huxley, *Man's Place in Nature,* 144; Huxley, "What Are Species?" 409–11.

August, 1860, "the man is clearly crazy, but it is a capital dig into the parsons—it stirs up Creation and much good comes out of such thorough discussions."[23] The truth was, as William Stanton has observed, "Darwin had beaten him at his own game and outdone even Nott at infidelity."[24] "The Old Roman," as Nott was called in his last years, admitted to William Henry Anderson of the Medical College of Alabama that he would not have published *Types of Mankind* "if the prehistoric period of man had been so firmly established [as] when he was making his investigations."[25]

By 1866 Nott concluded that Darwin's theory had not really threatened much of what he had written concerning the races after all. Though he accepted Darwin's thesis as to man's basic unity, he saw nothing in the theory to suggest that the races of man, "if not distinct species, are at least *permanent varieties.*"

> The question then, as to the existence, and *permanence* of races, types, species, or permanent varieties, call them what you please, is no longer an open one. Forms that have been permanent for several thousand years, must remain so at least during the lifetime of a nation. It is true, there is a school of Naturalists among whom are numbered the great names of Lamarck, Geoffroy Saint-Hilaire, Darwin and others, which advocates the *development* theory, and contend not only that one type may be transformed into another, but that man himself is nothing more than a developed worm; but this school requires *millions of years* to carry out the changes by infinitesimal steps of progression.[26]

Each permanent race type had a peculiar "physique" or anatomical structure and a kindred "moral" or instinct that was inseparable. Though physique and moral might change through millions of years, they were valid distinctions for the varieties of men living in a nation's history. He concluded that Darwin's theory was irrelevant for the problems of immediate and temporal political affairs. Darwin's "refinements of science" had no connection with the permanent characteristics of Negro inferiority since "the

[23] Nott to Squier, in *Ephraim Squier Papers*, Library of Congress, MSS.
[24] Stanton, *The Leopard's Spots*, 185.
[25] William H. Anderson, *Biographical Sketch of Dr. Josiah C. Nott* (Mobile, Ala., 1877), 6–7; William M. Polk, "Josiah C. Nott," *American Journal of Obstetrics and Diseases of Women and Children*, LXVII (1913), 957–58.
[26] Nott, *Instincts of Races*, 4–5.

Freedmen's Bureau will not have vitality enough to see the negro experiment through many hundred generations, and to direct the imperfect plans of Providence."[27]

In 1866 the London *Anthropological Review* published an open letter written by Nott to Major General O. O. Howard, superintendent of the Freedmen's Bureau. Nott's alleged purpose in writing the letter was to show Howard, to whom Nott believed the Negro and slavery were only abstractions, "the physical and civil history of the negro race, that it is now, wherever found, just what it was 5000 years ago," and from such evidence "to inquire what position Providence has assigned it in the affairs of our world."[28] Combining the disciplines of history and science, Nott chose arguments that naturalists would use equally upon the Indian, Negro, and Chinese later in the nineteenth century.

Nott's argument with Howard was essentially that the Freedmen's Bureau was preventing progress in the United States and the South in particular by attempting to place the Negro in full equality with the white population. Though slavery had been a means of developing the resources of the South, he admitted that the peculiar institution had become "a great and growing evil." Up until the present, Nott argued, "the history of the negro race is simply a page of natural history—it has no intellectual history, because God has not endowed it with the faculties necessary to preserve written records." In his natural state, that designed by God, the Negro was more pious, moral, honest, and useful than in the "unnatural state" which the bureau was attempting to achieve by educating him. The Negro found it far easier to learn the vices of white society than to absorb the useful and intellectual virtues of the more progressive Caucasian. The whole intellectual and social system of the country would improve with the substitution of whites for Negroes in the development of the nation.[29]

There was nothing in history to prove that the Negro had been anything more than a slave and laborer for thousands of years. In

[27] *Ibid.*, 4.

[28] Josiah C. Nott, "The Negro Race," *Anthropological Review*, IV (July, 1866), 103; Nott, "The Problem of the Black Races," *De Bow's Review*, n.s., I (Mar., 1866), 266–83.

[29] Nott, "The Negro Race," 105–6.

the earlier records of man the Egyptians depicted the Negroes as slaves; neither their features nor their position had changed through subsequent generations. Arguments for the progressive development of the Negro race were without foundation. In 4,000 years marked by successive progression by the other races of man, the intellect of the Negro "has been as dark as his skin, and all attempts in and out of Africa have failed to enlighten or develop it beyond the grade for which the Creator intended it."[30]

When reformers spoke of the Negro's intellectual abilities, Nott rejoined by saying that such individuals usually used the mulatto and not the pure-blood as the example. Nott argued that such examples were not valid. He was equally opposed to assertions of the intellectual abilities of Frederick Douglass. He was, said Nott, "the most brilliant mulatto now before the public, and he is nothing more than what St. Paul calls a 'pestilent fellow.'" "He has just brains enough," he wrote, "to talk fluently about matters he does not comprehend, and to spit out the venom of a blackguard." History afforded no better example of mulatto ability than in the failure of the Negroes of Haiti. The mulatto caste that ruled the island "swept every remnant of civilization from the country, which soon relapsed into savageism." When the white and black races mixed, he wrote, they produced a variety that was both physically and intellectually intermediate between the two original stocks. "They are more intelligent than the blacks, and less so than the whites," Nott argued. He doubted, however, if the added intellect was enough "to improve [them] to any useful degree." On the other hand "it is certain that the white race is deteriorated by every drop of black blood infiltrated into it—just as surely as the blood of the greyhound or pointer is polluted by that of a cur."[31]

In reference to those who argued the Negro's inferiority from environmental isolation, Nott rejoined that the Negro's position was no different from the isolation of the Russian Empire, cut off for centuries from the rest of civilization. But the Russians, "opposed by every obstacle that could obstruct the progress of a people

[30] *Ibid.*, 106–7.
[31] *Ibid.*, 111.

. . . triumphed speedily and nobly." The difference between the Russians and the black races was one of cranial capacity rather than environmental isolation. In order to substantiate his assertions, Nott suggested a study of Samuel Morton's measurements of brains:[32]

BRAINS IN CUBIC INCHES	
Teutonic	92
Pelasgic	84
Celtic	87
Semitic	89
Ancient Pelasgic	88
Malay	85
Chinese	82
African	83
"Hindostanee"	80
Fellah (Modern Egyptians)	80
Egyptian (Ancient)	80
Toltecan family	77
Barbarous tribes	84
Hottentot	75
Australian	75

Nott rationalized the apparent difficulty in the fact that the African had a larger brain size than the Chinese and a similar brain size to the Malay and Hindustani by explaining that in "the negro the posterior or animal part of the brain greatly preponderates over the anterior or intellectual lobes." To deny the relevancy of the evidence, he argued, was to deny history. Though the Alabama scientist was not a complete believer in the assertions of phrenology, he did accept as fact that there were divisions in the brain and that the intellectual faculties of man were grouped in the front of the brain. "Who will deny," he wrote, "the broad historical fact, that the white, which are the large-brained races, have governed the world from time immemorial, and have been depositories of true civilization?" The table of race-brain measurements showed a

[32] *Ibid.*, 112–13. The tabulation comes from Nott and George R. Gliddon, *Types of Mankind* (Philadelphia, 1854), 454.

"sliding scale of seventeen cubic inches of brain between the Hottentot and Australian at one extreme, and the Teutonic races at the other."[33]

History, according to Nott, proved that the nearest approach the Negro made to civilization had been in some sort of subordinate position to the white race and that, when left alone, he reverted to savagery. False philanthropy could enlarge neither brain capacity nor intellectual ability. There was no evidence in past history to suppose that through education, implanted in successive generations, the brain of the Hottentot would enlarge itself some 17 cubic inches to equal that of the Anglo-Saxon. There was no evidence from history that the Negro had in any way developed beyond the intellectual ability he had at the time of Egyptian civilization. Given the opportunity of countless civilizations, the Negro had remained inferior. By the same token, the skull sizes of present English nobles had not changed from those of ancient Britons. Education had achieved nothing toward enlarging the brain or expanding the intellect of Britons—the intellect was always there, ready to respond to civilization, just as Russian serfs were now responding. For the Negro races, however, brain capacity was too low, and no amount of environmental change could bring about a responding intellectual development. The stimulation of environment affected only those brains which contained the innate capacity for progressive response.[34]

As a result of brain deficiency, the Negro race was dependent for its perpetuation on the kindness and interest of a superior race. The Civil War forever destroyed the attention given to the Negro by the southern plantation owner, a situation which now meant doom for the race in America. With all the evidence of history against the Negro, Nott concluded that the Negro's present antipathy for labor meant that he was doomed to extinction. Driven by his own instincts and unwillingness to work into the towns of the United States, the Negro, according to Nott, would die out.[35]

In similar language Harvard zoologist Louis Agassiz suggested

[33] Nott, "The Negro Race," 113.
[34] *Ibid.*, 114.
[35] *Ibid.*, 116.

that the government have two separate policies in order to contend with the half-breed and the pure black. He believed that the government should offer the black race every possible chance to secure "the fullest developments of its capabilities." He felt that the pure Negro should remain in the South where he could live a life commensurate with his physical abilities and mental aptitude. The South was most suited to his physique and there was no reason to believe that the pure black would die out.[36] Those who moved north, however, would linger for a while but eventually die out, since the northern climate was totally unsuited to the Negro's character. The mulatto's existence, he argued, would most likely be only transitory. His physical disabilities, despite an increased mental aptitude, made him an artificial "hybrid" that would die out. Legislation, therefore, should be designed "to accelerate [his] disappearance from the Northern States," while real substantive rights should pertain only to the pure black.[37]

"I beseech you," wrote Agassiz to Samuel G. Howe of the Sanitary Commission, "to allow no preconceived view, no favorite schemes, no immediate object, to bias your judgment and mislead you."[38] Agassiz had favored emancipation from a philanthropic, physiological, and ethnographic point of view. There was no more malicious practice than slavery, except perhaps the doctrine that all men were equal, "in the sense of being equally capable of fostering human progress and advancing civilization."[39] There was no substance to the belief that the condition of the Negro was wholly due to slavery. Such a belief obscured some 4,000 years of past Negro history. The Negro was entitled to freedom, but in no way was he "capable of living on a footing of social equality with the whites in one and the same community without becoming an element of social disorder." White society, Agassiz wrote, should parcel out rights to pure blacks in "successive installments," in proportion to their ability and capacity to take on responsibility.[40]

[36] Agassiz to Samuel Gridley Howe, Aug. 9, 1863, in Agassiz, ed., *Louis Agassiz,* 599–600.

[37] Agassiz to Howe, Aug. 10, 1863, in *ibid.*, 608.

[38] *Ibid.*, 602.

[39] *Ibid.*, 604.

[40] *Ibid.*, 607.

"Let us beware," he warned, "of granting too much to the negro race in the beginning lest it become necessary hereafter to deprive them of some of the principles which they may use to their own and our detriment."[41]

DARWIN AND *The Descent of Man*

It was not until 1871, with his *Descent of Man,* that Darwin chose to carry out what, to many, appeared to be logical corollaries of his original argument in *Origin of Species.* In a certain sense *The Descent of Man* was a restatement of the theories of those "Darwinissimists" who had first exploited Darwin's thesis and applied it to man. Darwin found the similarities between man and the lower animals too great to conceal. "The world," he wrote, "appears as if it had long been preparing for the advent of man."[42] He hoped that when both monogenists and polygenists accepted the principle of evolution, the dispute would "die a silent and unobserved death."[43] That races existed, there could be no doubt, but to argue that they were distinct species was obscuring the fact "that they graduate into each other, independently in many cases . . . of their having intercrossed."[44] Darwin drew his conclusions from the evidence accumulated by Paul Broca, Charles Lyell, and the American naturalist John Bachman, who showed that the races, when crossed, were quite fertile. But aside from the relationship of all races to each other, mental and physical characteristics were distinct "in their emotional, but partly in their intellectual faculties."[45]

Darwin accepted the fact that intellectual faculties differed among the various races of men. He read the conclusions of Benjamin A. Gould's somatometry on the northern army during the Civil War and the measurements of Dr. J. Barnard Davis, who

[41] *Ibid.,* 608.
[42] Charles Darwin, *The Descent of Man and Selection in Relation to Sex* (New York, 1927), 169.
[43] *Ibid.,* 184.
[44] *Ibid.,* 170, 178.
[45] *Ibid.,* 171, 174–75.

explained differences in internal capacity of brains of Europeans, Americans, Asiatics, and Australians.[46] "The American aborigines, Negroes and Europeans," Darwin wrote, "are as different from each other in mind as any three races that can be named."[47] Yet, despite these differences, the highest as well as the lowest races were connected by fine gradations. It was entirely possible, he believed, that race varieties "might pass and be developed from each other."[48] Such development, however, did not preclude the polygenist idea of development from the catarhine and platyrhine apes. Races did not necessarily form one single, monogenist, ascending series but, rather, a series of parallel lines linked somehow with "one extremely ancient progenitor." Thus the dolichocephales of Europe and Africa might have been the "cousin-german" of the dolichocephalic chimpanzee and gorilla of Guinea and the brachycephalic orangs of Sumatra and Borneo, while "the ancestor common to them both is farther off still."[49] Darwin utilized physiological structure in determining differences between the several so-called races. He explained the phenomenon of races as various human types that "remained distinct for a long period." In such cases, the varieties might just as well be called species. "Even a slight degree of sterility between any two forms when first crossed, or in their offspring, is generally considered a decisive test of their specific distinctness; and their continued persistence without blending within the same area, is usually accepted as sufficient evidence, either of some degree of mutual sterility, or in the case of animals of some mutual repugnance to pairing."[50]

The term "species" remained a plague for the anthropologist. Those who did not accept evolution looked upon the term as characterizing separate creations. Those who accepted evolution, on

[46] *Ibid.*, 171; Benjamin A. Gould, *Investigations in the Military and Anthropological Statistics of American Soldiers* (New York, 1869); J. Barnard Davis, "Contributions toward Determining the Weight of the Brain in the Different Races of Man," Royal Society of London, *Proceedings*, XVI (1867–1868), 236–41; Davis, "On the Weight of the Brain of the Negro," *Anthropological Review*, VII (1869), 190–92.

[47] Darwin, *Descent of Man and Selection in Relation to Sex*, 182.

[48] *Ibid.*, 66, 178.

[49] Topinard, *Anthropology*, 531; Darwin, *Descent of Man and Selection in Relation to Sex*, 156–58.

[50] Darwin, *Descent of Man and Selection in Relation to Sex*, 169–70.

the other hand, excused themselves from defining the term. They accepted that all races were descended from a single stock, "whether or not they may think fit to designate the races as distinct species, for the sake of expressing their amount of difference." Darwin felt that the term was arbitrary. It was so ill defined that "such early races would perhaps have been ranked by some naturalists as distinct species, if their differences, although extremely slight, had been more constant than they are at present, and had not graduated into each other."[51]

ALEXANDER WINCHELL

Gradually anthropologists grew more aware of the futility of the origination argument; "race" and "species" became arbitrary terms whose proof of existence lay too deep in the paleontological past. But while the feud itself died, those concepts of racial inferiority that existed as part of the origination feud became post-Darwinian vocabulary nonetheless. The so-called "inferior races" remained at the basis of evolutionary discussion. Geologist Alexander Winchell (1824–1891), a seventh-generation product of New England stock, epitomized in a certain sense the post-Darwinian confusion of race discussion. A graduate of Wesleyan University of Connecticut in 1847, Winchell began a teaching career in science that took him to the presidency of the Masonic University at Selma, Alabama, the opening of the Mesopotamia Female Seminary in Eutaw, Alabama, a chair of physics and civil engineering at the University of Michigan, the chancellorship of Syracuse University, a professorship of geology and zoology at Vanderbilt University, and finally back to Ann Arbor and a chair in geology and paleontology. Between 1862 and 1864 he wrote sixteen articles entitled "Voices from Nature" in the *Ladies' Repository* and spread far and wide a typological and ontogenetic sentiment that was essentially religious, rooted in the philosophy of Plato, dramatizing the "consummation of organic exaltation" in the human form and intelli-

[51] *Ibid.*, 180.

gence.[52] To read Winchell is to recapture the exhilarated mood of America's Transcendentalist generation.

We behold the long barred doors of nature, opening to admit us to her inmost shrines. We enter her sacred temple, and every object breathes the presence of the Infinite Mind. Every stone is inscribed with a word of revelation; and we ponder the meaning of these divine records, we feel thrilled with the conviction, that we have possessed ourselves of thoughts that were conceived in the mind of the Omniscient. We go forth from our communings, purified and exalted in soul; and instead of banishing Deity from the universe, we delight to know that he exists on every side of us.[53]

Yet Winchell's opinion changed in the 1870's and, for reasons still unclear, he accepted the evidence of Darwin. But, like many postwar American naturalists, Winchell became concerned with transposing Darwinism into "soft" teleology—an effort to bring evolution within the confines of design from a beneficent Maker. "Natural Selection seemed to be put in the place of the Athanasian creed," wrote Henry Adams. "It seemed a form of religious hope; a promise of ultimate perfection."[54] Winchell saw in the vital principle (variation) a proof of the greater power and wisdom of the Creator, preserving the economy of nature through invariable and inflexible laws. He saw design and symmetry in nature in the mold of external and physical causation. In effect, he set aside Darwin's dysteleology for a transcendental purposiveness, a nonselectional evolution closer to the monistic, deductive philosophy of Herbert Spencer. Winchell, like most early American evolutionists, never really weighed Darwin's words. "We were still in the Twilight of the Gods," wrote the ex-southerner and Unitarian Moncure Conway, "reverently spelt nature with a big N, and saw our goddess ever at her loom, but weaving with swift shuttles."[55]

There was yet another side of Alexander Winchell which was

[52] Alexander Winchell, *Sketches of Creation: A Popular View of Some of the Grand Conclusions of the Sciences in Reference to the History of Matter and Life* (New York, 1870), 378.

[53] Alexander Winchell, *Voices from Nature* (Ann Arbor, Mich., 1858), 3–4.

[54] Henry Adams, *The Education of Henry Adams* (New York, 1918), 231.

[55] Moncure D. Conway, *Autobiography, Memories and Experiences*, 2 vols. (Boston, 1904), I, 282.

grounded in the anthropological researches of Josiah Nott, George Gliddon, Paul Topinard, James Cowles Prichard, Alfred Wallace, Armand de Quatrefages, Paul Peschel, and the travel accounts of Barth and Lichtenstein. Perhaps it was not so much another side of Winchell as it was an effort to combine the formal aspects of science with biblical narrative. In 1880 Winchell wrote his famous *Preadamites;* it went through five editions before 1890. His belief in the existence of pre-Adamites incurred the displeasure of Vanderbilt University. "Such views," he was told, "are contrary to the plan of redemption" and he was asked to resign his chair.[56] But though Winchell "removed the incredibility of that doctrine as grounded in the descent of Negroes and Australians from Noah and Adam," he did maintain the essential unity of mankind.[57] The blood of the first human flowed in all men and therefore all were subject to God's redemptive grace. By pre-Adamitism he meant not the plurality of origins but simply that Adam had descended from the black race and not the black race from Adam. "Those who hold that the White race, the consummate flower of the tree, has served as the root from which all inferior races have ramified," he wrote, "may select their own method of rearing a tree with its roots in the air and its blossoms in the ground. I shall put the tree in its normal position." The Adamic stock, whose ethnological characters resembled those of the present Mediterranean race, derived from a "humbler human type." There was nonetheless a common consanguinity between the Adamite and the black races that accounted for an essential brotherhood and a common destiny.[58]

The brown and black races had constituted a large population in both Asia and Europe prior to Adam. Although Winchell continually affirmed their rights and responsibilities as members of humanity, he could not ignore "the ethnic chasm" which divided "them from the mass of Noachite humanity." Enclosed in the "bosoms of vast and impenetrable continents," it seemed as though

[56] William S. Studley, *Memorial to Alexander Winchell* (Ann Arbor, Mich., 1891), 16.

[57] Winchell, *Preadamites,* v.

[58] *Ibid.,* 297. For reaction to Winchell's book, see Edward Wilmot Blyden, president of Liberia College, *The Aims and Methods of a Liberal Education for the Africans* (Cambridge, 1882), 12–15.

nature "had contented herself to herd [the brown and black races into] regions where they would never mingle in the stir and strife of social and national struggles." To study the Negro past was not unlike an investigation into the natural history of a pig or horse, since the Negro wrote no history and achieved no "results for history to record." "Their thousands of years outlived are silent," he wrote; "not an echo of a former generation comes down to our apprehension." The gap that separated brutishness, inertia, indolence, and stupidity from the "indomitable energy, the flashing intellect, and the heaven-reaching aspirations which have made our planet the abode of civilization" covered much "more than a few centuries . . . and must find its origin deep in the ages, and in the early divarication [*sic*] of courses of events which have emerged in our own times."[59]

In order to distinguish pre-Adamites from Noachites, Winchell borrowed profusely from nineteenth-century methods of race classification. Prognathism marked the Noachites with a higher facial angle, while the gradation moved down through the Mongoloids to the lowly Bushmen, whose cranial index was "extreme and even frightful."[60] He suggested, as did Friedrich Müller and Darwin, that the Hottentots and the Ainos of Japan were a "racial ruin," the perishing remnants of one of the isolated roots in the pre-Adamic stock.[61] The Negro, Winchell argued, was structurally inferior. He was not the result of structural degradation or degeneration from the Caucasian, since Winchell saw no instance of this phenomenon ever occurring in nature. "I hold it to be the edict of Nature," he wrote, "that no type of organization, having once entered the portals of a higher life, shall be permitted to retreat." The Negro "is the best he has ever been," and therefore was not descended from Adam but, rather, Adam from him.[62] He emphasized that structural deterioration was impossible and wholly distinct from the evidence of cultural deterioration. Hence the Portuguese in Malacca, breathing the influence of an inhospitable environment,

59 Winchell, *Preadamites,* 156–57.

60 *Ibid.,* 171.

61 *Ibid.,* 86; Darwin, *Descent of Man and Selection in Relation to Sex,* 181–92.

62 Winchell, *Preadamites,* 275–76.

brought about a deadening of his civilization. But in no instance did cultural degradation bring about a corresponding physical shrinking of the cranium, a dolichocephalous form, or any other structural reversion. "Never, except as inherited," he wrote, "does Negroid prognathism develop, or the arm or the heel lengthen, or the pelvis become more oblique."[63]

After establishing what he felt was the fallacy of structural degeneration, Winchell set out to explain those elements of structural inferiority basic to Negro or pre-Adamite. The Negro skull was extremely thick "and is often used for butting, as is the custom of rams." Flattened on the top, it was well suited for carrying burdens. His clavicle was larger in proportion to the humerus and therefore approached the structural organization of the ape. His scapula was shorter and broader, his pelvis narrower than even the yellow races and was, in fact, inclined like the anthropoid.[64] The Negro brain was darker and "its density and texture . . . inferior"; brain convolutions were far simpler. Winchell, along with Agassiz and Marcel de Serres, thought the Caucasian brain during its embryonic development presented "in succession the conformations seen in the Negro, the Malay, the American and the Caucasian," while the Negro brain was arrested.[65]

Winchell warned the Caucasian of his responsibilities toward the inferior races. "I am responsible," he wrote, "if I grant him privileges which he can only pervert to his detriment and mine; or impose upon him duties which he is incompetent to perform or even to understand."[66] With this admonition Winchell protested against the miscegenation suggestions of Wendell Phillips, Bishop Gilbert Haven, Canon George Rawlinson, and David Croly. The interfusion of Noachite and pre-Adamite would cause irreparable harm to the nation. Turning to Sanford B. Hunt's study of the United States Sanitary Commission medical examinations made during the Civil War, he concluded that where white blood predominated

[63] *Ibid.*, 280.

[64] *Ibid.*, 171–72. Many of his conclusions were taken from Jeffries Wyman and Thomas S. Savage, "Troglodytes Niger," *Boston Journal of Natural History*, IV (1843), 1–24.

[65] Winchell, *Preadamites*, 250–51.

[66] *Ibid.*, 266.

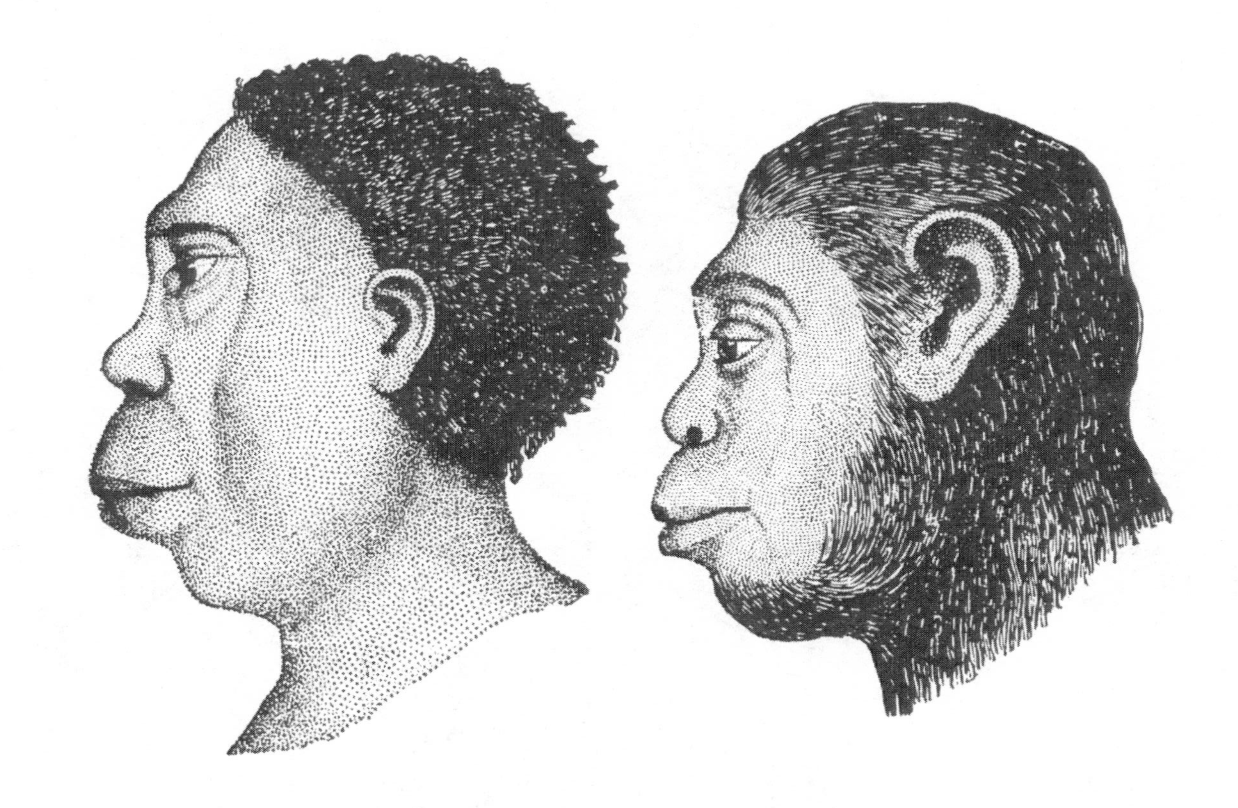

Comparison of female Hottentot with female gorilla (from Alexander Winchell, *Preadamites* [1888]).

Courtesy of Indiana University

in mixed breeds, it "exercised a preponderating influence in favor of cerebral development." However, "the inverse predominance of Negro blood" left the brain "in a condition of inferiority approaching even that of the pure Negro." Similarly, Winchell looked to Benjamin A. Gould's Civil War anthropometric statistics on lung capacity and chest circumference as corroborating the evidence of Hunt's brain analysis.[67]

Winchell, like other scientists and naturalists who bridged the era of Darwin's initial impact, saw little in his thesis to discredit or disparage his own concepts of race. Louis Agassiz, who opposed Darwinian evolution, and Nott and Winchell, who accepted it although on different grounds, illustrate how easily the views of the older pre-Darwinian concepts of racial inferiority remained essentially the same in the post-Darwinian period. Darwinism, however it might change the development concepts of the newer scientific generation, remained initially neutral in the race vocabulary of its successors. In time the hypothesis of evolution and the factors of variation and survival of the fittest gave added scientific sophistication to the heritage of the naturalist's racial characterizations. By then, however, the age brought newer leaders to the fore in industry and enterprise who wanted their qualities reflected in the more contemporary definition of racial superiority. Yet while the national tastes led to a newer vocabulary, the race characterization and stereotypes of the older generation seemed in no way to conflict with those concepts of racial inferiority that captivated the younger generation. Evolutionary vocabulary reflected the country that bred it and, in reflecting it, wore the prejudices of the land that gave it birth.

[67] *Ibid.*, 83.

IV *Race and the Concept of Progress in Nineteenth-Century Ethnology*

ONE OF THE EARMARKS of change in pre-academic anthropology in late nineteenth-century America was the separation of physical anthropology from cultural anthropology, then represented by ethnology. Despite their separation, each utilized the work of the other in an attempt to find the various missing links in both anatomical and cultural development. This enabled both physical and cultural anthropology to aid the other by borrowing "links in one chain of events to supply the gaps in another."[1] Did not the existing nations and tribes, each with their particular habits, customs, and institutions, compare to the taxonomic series? Did not societal development from savagery to civilization mark embryonic or ontogenetic series? Were not the crude implements

[1] James G. Frazer, *The Scope of Social Anthropology* (London, 1908), 18; Felix M. Keesing, *Cultural Anthropology: The Science of Custom* (New York, 1958), chap. I.

of stone, bronze, and iron developed through historical periods representative of geological or phylogenetic series? And, last, did not the conditions of peace, prosperity, and anarchy reveal a corresponding pathological series in human development? Science in the late nineteenth century became suffused with the language of evolution and although Darwin had excluded teleology from his evolutionary scheme, at the popular level both scientists and social scientists continued to espouse teleology under the guise of inevitable cultural progress.[2] Concepts of racial inferiority that existed in the vocabulary of either physical or cultural anthropology assimilated easily into the language of the other. Differences which existed within similar culture grades were obscured by efforts to integrate all observations into a unilinear concept of cultural evolution.[3]

Ethnology, as understood by nineteenth-century anthropologists, limited its investigations to the rudimentary beginnings of human society. Essentially, ethnology was the comparative and developmental study of social man and his culture. Concerning himself with the science of culture, the ethnologist enumerated the conditions and modes of existence of specific nonwestern peoples and only touched tangentially upon the contemporary problems of western life. Although he was not studying his own society, he made liberal use of his studies of human thought and institutions in their embryological stage to suggest the same unlinear phylogeny for advanced civilizations. The ethnologist thus walked an unstable course between his science and his assumptions in the nineteenth century, offering suggestions in and out of his discipline and generalizing about human behavior in its various aspects. Every man, wrote Otis T. Mason, was his own ethnologist. There was neither a priesthood nor a laity in the discipline. Each man

[2] Joseph LeConte, "Scientific Relation of Sociology to Biology," *Popular Science Monthly*, XIV (Feb., 1879), 427.

[3] Melville Herskovits, "Some Problems of Method in Ethnology," in Robert F. Spencer, ed., *Method and Perspective in Anthropology* (Minneapolis, 1954), 5–6; John R. Swanton, "Some Anthropological Misconceptions," *American Anthropologist*, n.s., XIX (Oct.-Dec., 1917), 459–70; Franz Boas, "Some Recent Criticisms of Physical Anthropology," *ibid.*, I (Jan., 1899), 98–106.

was "the investigator and the investigated,—the judge, the jury, and the prisoner at the bar."[4]

The ethnologist or cultural evolutionist was more vulnerable in his use of the comparative method than the physical scientist. His archeological foundations were further removed from and weaker than the paleontological foundations of the biological evolutionist. "Cultural evolutionism had no embryology to support its argument," wrote Erwin Ackerknecht, yet the cultural evolutionist in the nineteenth century assumed a parallelism between biological and cultural evolution. The physical anthropologist accepted the parallelism, but his own work on the physical aspects of man gave his research an empirical foundation so that he was not dependent on the generalizations of the cultural anthropologist. The cultural anthropologist, however, drawing upon his sketchy ethnological accumulations, continued on his perilous course of supra-organic analogy.[5]

For those early cultural evolutionists like Herbert Spencer, Sir James G. Frazer, Sir Edward B. Tylor, Lewis Henry Morgan, Edward Westermarck, and Hutton Webster, cultural evolution was but a chapter of biology itself. Just as biology suggested a sequence of forms ascending from homogeneity to heterogeneity, from the single-celled to the multiple-celled organism, so the cultural anthropologist described the races of mankind moving through successive orders of complexity. Similarly, just as organic evolution became essentially connected with the process of heredity, the cultural evolutionist characterized social evolution and the

[4] Otis T. Mason, "What Is Anthropology?" in Anthropological and Biological Societies of Washington, *Saturday Lectures* (Washington, D.C., 1882), 42.

[5] Erwin Ackerknecht, "On the Comparative Method in Anthropology," in Spencer, ed., *Method and Perspective in Anthropology,* 122; Frazer, *Scope of Social Anthropology,* 18; Simon N. Patten, "The Failure of Biologic Sociology," American Academy of Political and Social Science, *Annals,* IV (May, 1894), 917–47; George E. Fellows, "The Relation of Anthropology to the Study of History," *American Journal of Sociology,* I (July, 1895), 41–49; Frederick J. Teggart, "Prolegomena to History: The Relation of History to Literature, Philosophy, and Science," University of California, *Publications in History,* IV (1916), 268–69; Alexander Goldenweiser, *History, Psychology, and Culture* (New York, 1933), 125–26; Joseph Jastrow, "The Natural History of Analogy," American Association for the Advancement of Science, *Proceedings,* XL (1892), 336, 352.

process of civilization's advance in the same hereditary schema.[6] In his study of a given race or people the cultural evolutionist was "guided far more by its dead than by its living." Accordingly, he chose to judge a people by their past: "however much its ancient elements are no longer living as such, they nevertheless form its trunk and body, around which the live sap-wood of the day is only shell and surface."[7]

In accordance with his emphasis on the relations of all living and dead things to one another, the ethnologist erected no boundary between historic and prehistoric time or between historic and unhistoric peoples. Both the physiological or purely biological structure of man and the aspects of his social life became part of the same cosmic development. Obscured by the comparative method of analogy, explanations of biological and social evolution became synonymous in meaning. Furthermore, what distinctions existed between the physical scientist's evidence of morphological evolution and the teleological implications of the century's belief in progress merged into the ethnologist's supra-organic schema of development. In a sense the ethnologist's concept of social evolution grew not only out of the biological theory of descent but also out of his belief in progress, which was but a value judgment projected into a scientific process. Though the ethnologist of the nineteenth century relied on the word "evolution" more than the word "progress" to define his theory of culture, he actually used "evolution" to mean "progress." Evolution, or the theory of descent, implied for him a teleological projection that assured the perfectability of man through natural selection.[8]

Ethnologists like neo-Lamarckians John Wesley Powell and W J McGee of the Bureau of American Ethnology made great efforts to clarify the nature of human society. "The course of human

[6] Alfred L. Kroeber, "The Superorganic," *American Anthropologist*, n.s., XIX (Apr.-June, 1917), 167; I. L. Murphree, "The Evolutionary Anthropologists: The Progress of Mankind. The Concepts of Progress and Culture in the Thought of John Lubbock, Edward B. Tylor, and Lewis H. Morgan," American Philosophical Society, *Proceedings*, CI (1961), 265–300.

[7] Kroeber, "The Superorganic," 186.

[8] Teggart, "Prolegomena to History," 243–47; Joseph LeConte, "Evolution and Human Progress," *Open Court*, V (Apr., 1891), 2779–83; LeConte, "The Test of Progress," *ibid.*, (Aug., 1891), 2915–16.

events is not an eternal round," wrote Powell. Far from mere repetition, he argued, "there is always some observable change in the direction of progress."[9] McGee wrote that "each generation is a little better than the one that went before, on the average . . . and that consequently the trend of human development is an upward trend."[10] Recent students of ethnology have argued that American ethnological theorizing over the term "progress" was a reflection of the earlier writings of men like Francis Hutcheson, Thomas Reid, Adam Ferguson, Lord Kames, and William Robertson of the Scottish school of common sense. Like the Scottish school, American ethnologists attempted to create a sociology of progress which transformed Christian millennial theology "into a certainty of the God-ordained, intelligent self-sufficiency of modern man to work out his own way in his common sense, his analytic reason, and his specific moral sense."[11]

Yet while the Scottish influence may have been evident in American ethnology, it appears clear that ethnologists in the late nineteenth century belonged to no formal school of philosophy. Indeed, they used the concept of progress with little regard for specific definition. For most, it meant little more than "evolution" or "a march onward."[12] Similarly, the criteria for progress varied greatly. For Henry Bates and for Frank Baker, editor of the *American Anthropologist* from 1891 to 1898, progress concerned only those races whose evolution was unobstructed, whose cranial sutures were still "plastic," and whose brain weight and prognathism evidenced a development away from quadrumanous features.[13]

[9] John Wesley Powell, "From Barbarian to Civilization," *American Anthropologist,* o.s., I (Apr., 1888), 97.

[10] W J McGee, speech, Aug. 26, 1904, Library of Congress, MSS, fol. 27.

[11] Roy Harvey Pearce, *The Savages of America: A Study of the Indian and the Idea of Civilization* (Baltimore, 1956), 82; Gladys Bryson, *Man and Society: The Scottish Inquiry of the Eighteenth Century* (Princeton, N.J., 1945), 36, 41–42, chap. IV; David Bidney, "The Idea of the Savage in North American Ethno-History," *Journal of the History of Ideas,* XV (1954), 322–27; A. R. Radcliffe-Brown, "Evolution, Social or Cultural," *American Anthropologist,* n.s., XLIX (Jan.-Mar., 1947), 78; John Wesley Powell, "Darwin's Contributions to Philosophy," Washington Biological Society, *Proceedings,* I (May, 1882), 66.

[12] LeConte, "The Test of Progress," 2915.

[13] Henry Bates, "Discontinuities in Nature's Methods," *American Anthropologist,* o.s., I (Apr., 1888), 135–46; Frank Baker, "The Ascent of Man," *ibid.,* III (Oct., 1890), 297–319.

John Wesley Powell and Lewis Henry Morgan, although they spoke optimistically of progress for all peoples, actually limited the full meaning of the term to only those peoples whose race history clearly evidenced a movement out of savagery and barbarism into civilization. The American Indian, who had not yet developed an agricultural society, possessed no "progressive spirit." The Indian's position in the hunter stage placed him at "the zero of human society," from which "there was no hope of elevation."[14] For W J McGee, progress meant the capacity of races to transcend blind natural forces through purposeful action, and as an evidence of progress he decided upon the progression toward or regression from democracy as a useful guideline.[15]

Contributors to the *American Anthropologist* in the late nineteenth century—historian and explorer Adolph F. Bandelier (1840–1914), anthropologists Henry Bates and Daniel G. Brinton, historian of medicine Frank Baker (1841–1918), curator of ethnology of the Smithsonian Institution, Otis T. Mason (1838–1908), and ethnologists W J McGee and John Wesley Powell—took the concept of progress for granted in their writings. Their primary concern was in defining the stages of human progress and analyzing the processes involved in each of the stages. Discussion, therefore, concerned factors in cultural development that reflected elements of natural selection, "man's own nationality or emotional nature, the characteristics of particular environments, or the diffusion of specific cultural innovations."[16]

Henry Bates, like so many late nineteenth-century ethnologists, assumed that cultural development involved a corresponding brain development in the race. Borrowing his theory of human develop-

[14] Lewis Henry Morgan, *League of the Ho-dé-no-sau-nee, or Iroquois* (Rochester, N.Y., 1851), 141–43; Morgan, *Ancient Society: Or Researches in the Lines of Human Progress from Savagery through Barbarism to Civilization* (New York, 1877), vi, viii, 41–42; William C. Darrah, *Powell of the Colorado* (Princeton, N.J., 1951), 282–83; John Wesley Powell, "Sociology, or the Science of Institutions," *American Anthropologist*, n.s., I (Oct., 1899), 724–28.

[15] W J McGee, "The Trend of Human Progress," *American Anthropologist*, n.s., I (July, 1899), 401–7; McGee, "The Citizen," *ibid.*, o.s., VII (Oct., 1894), 352–57.

[16] Frederica de Laguna, "The Methods and Theory of Ethnology," in Laguna, ed., *Selected Papers from the American Anthropologist, 1888–1920* (New York, 1960), 784.

ment from both Spencer and Powell, he saw an advancement from brutish militancy where man, like the animals, engaged in struggle for existence, to later stages of industrialism and the peaceful arts. Accompanying the later stages of growth was a corresponding moral improvement. Agreeing with Spencer, Bates thought that the development of ethics was "indispensable to the social condition" of advanced races. Evolution from the brute stages of militancy brought growing amounts of leisure, the use of the imagination, the arts of peace, and an increasing usage of the inventive faculties.[17]

In his "Discontinuity in Nature's Methods," written for the *American Anthropologist* in 1888, Bates saw cultural development as a Spencerian biological evolution where psychological development and cultural achievement were directly related to each other. As the inventive faculties freed man from "slavish toil" and the exigencies of self-protection, man "conjoined alien organs [i.e., tools] with his structure."[18]

> The development of the inventive faculty, as the distinguishing characteristic of mind, caused a modification of the old plan of progress by selective extermination. . . . Henceforth, natural selection affected only mental and ethnic qualities, through modification of his nervous structure. Instead of developing specialized organs, he began to construct extraneous ones for his use, having arrived at the specialized hand, by which such a new departure became possible. The discontinuity which especially characterizes man's development after this stage is his mental in place of physical evolution, coupled with evolution by extraneous organs.[19]

The brain of man in accelerated evolution was able to reach out, utilize, originate, construct, improve, and reproduce creations to replace restrictive physiological activities. Bates humorously suggested the brain's independence of the body's organs in the anecdote of the English army veteran who, before the eyes of his astonished coolie servant, "kicked off his right leg, detached an

[17] Bates, "Discontinuities," 144.
[18] *Ibid.*, 138–39.
[19] *Ibid.*, 135–36.

arm, deposited one eye in a glass of water, removed the upper and lower teeth," removed his wig, and then requested his paralyzed attendant "to unscrew his head."[20]

Geologist and anthropologist W J McGee (1853–1912) saw advancements through culture gradients from savagery to civilization as indicative of a corresponding cranial development.[21] Though he spent most of his public service as ethnologist in charge of the Bureau of American Ethnology, McGee participated at various times with the United States Geological Survey, the Louisiana Purchase Exposition in St. Louis, the St. Louis Public Museum, and the Inland Waterways Commission. He built upon the conclusions of Frank Baker and other American scientists interested in physical anthropology. Like Baker, he found evidence of increased osseous framework and brain capacity from the early *Pithecanthropus erectus* to enlightened man.[22] Likewise, he looked to the earlier studies of James Dwight Dana and Othniel C. Marsh on cephalization in subhuman forms as corroborating similiar phenomena in man. The transition in cephalization from the earthen graves of Europe's cavemen to the modern dissection rooms, from the "retreating type [of cranial conformation] of [George] Washington" to "the full-forehead type of the living statesman," told a story of progressive cranial capacity and "decrease among none." The records seemed to prove that cranial correlation with culture grade was so close "that the relative status of peoples and nations of the earth may be stated as justly in terms of brain-size as in any other way."[23]

Corresponding to Dana's earlier experiments on cephalization, McGee suggested the concept of *cheirization,* the coordination of the "initiative and directive faculties" of man. Like Bates, McGee believed that increased brain capacity advanced race ability far beyond physical development of other bodily parts—a situation

[20] *Ibid.*, 142.

[21] W J McGee, "The Science of Humanity," *American Anthropologist,* o.s., X (Aug., 1897), 241–72. McGee always signed his name "W J" and omitted any periods.

[22] Baker, "Ascent of Man," 297–319; Emma R. McGee, *Life of W J McGee* (Farley, Iowa, 1915), 69–70.

[23] McGee, "Trend of Human Progress," 409–10; Charles Schuehert, "O. C. Marsh," National Academy of Sciences, *Biographical Memoirs,* XX (1939), 1–78.

which brought about increased utility rather than physical change in the human organs. In the human hand, for example, somatological-psychological development externalized through "manifestations of manual dexterity among cultivated men." Similar processes involved modulations of voice, "eloquence of eye," "robustitude of limbs," and "sensitiveness of skin to touch and temperature."[24] The "sleepy eye," to which Dr. Samuel Morton had earlier referred in speaking of the backwardness of the American Indian, focused new meaning upon the words of these later ethnologists.[25] Though the term used by McGee to explain somatic development was new, the topic had long been a part of race study. Post–Civil War physician John H. Van Evrie had argued the futility of race improvement in the Negro on the basis of what McGee later called cheirization. In *White Supremacy and Negro Subordination* Van Evrie wrote that "the coarse, blunt, webbed fingers of the negress could not in any length of time or millions of years be brought to produce those delicate fabrics or work those exquisite embroideries which constitute the pursuits or make up the amusements of the Caucasian female." The "obtuseness of the sense of touch in the fingers," reflecting the limited capacity of the Negro intellect, relegated the race to the "grosser trades" which required "little more than muscular strength and industry to practice them."[26]

According to McGee, centrifugal or outward motions apparent in the physical dexterity of the higher cultures in remaking their secondary environment far surpassed and marked them off from the centripetal or inward movements of the primitive. Somatic changes were "charts to that highroad to human progress."[27] In other words, "the witnesses of somatic development from race to race, from antiquity to modernity, and from generation to generation are many and in the main consistent; the skull has risen from the simian type, the skeleton has become more upright and better

24 McGee, "Trend of Human Progress," 411; McGee, "The Seri Indians," Bureau of American Ethnology, *17th Annual Report* (1895–1899).

25 Samuel G. Morton, *An Enquiry into the Distinctive Characteristics of the Aboriginal Race of America* (Boston, 1842), 6.

26 John H. Van Evrie, *White Supremacy and Negro Subordination* (New York, 1868), 121.

27 McGee, "Trend of Human Progress," 411–12; McGee, *Anthropology at the Madison Meeting* (Washington, D.C., 1893), 440–41.

adjusted to brain-led activities, the muscles have gained and are still gaining in efficiency if not in absolute strength, the faculty for work (or normal exercise of function) is multiplied, the constitution is improved in vigor, life has grown longer and easier, and perfected man is over-spreading the world."[28]

McGee's conclusions by his own definition reflected an attempted synthesis of the writings of Darwin, Spencer, Lamarck, Bacon, and Powell. From Darwin and Lamarck he drew the idea that organisms interacted with the environment and that these efforts perpetuated in successive generations. From Spencer he extracted the idea that "organized bodies are composed of highly differentiated terrestrial substances combined in such manner as to perpetuate themselves through the continued maintenance of internal and external relations." Organisms formed a hierarchical table where the highly differentiated dominated the lesser organized, and the brain, a product of the highest degree of differentiation, was "the organ of the mind, [whose] function is the conservation and creation of intelligence." He drew from Bacon and John Wesley Powell the belief that mental development reflected the "directness or indirectness of its contact with nature." Hence, man's progression from the medium of muscles to created machines marked greater somatic differentiation as it signaled demotic or activital progression of culture.[29]

McGee saw no problem in discussing the ethnology and progress of all peoples as circumscribed within the same singular framework. Like scientist Daniel Brinton, he insisted that "any two minds must be expected to respond similarly to similar stimuli." From this point of departure, the "American Monroe doctrine of anthropology," as he called it, McGee insisted that minds, "wheresoever placed, must develop along essentially parallel or converging lines."[30] Human activities, irrespective of tribe, nation, or continent, "all diverge in form, yet converge in essential quality and in their effects on mankind."[31] Thus "it follows that, just as any

[28] McGee, "Trend of Human Progress," 414.
[29] *Ibid.*, 424–25.
[30] *Ibid.*, 427.
[31] *Ibid.*, 448.

two organisms of the same species are like in physiologic process and in response to external stimuli, so any two brains of equal faculty must function alike or so nearly alike as the environments by which their final shaping was given. Accordingly, the much-mooted unity of the human mind would appear to be nothing more than a manifestation of cerebral homology (itself a record of eons of organic development) perfected during the final eon of demotic process."[32]

In view of the earlier controversy concerning the origin of man, McGee contended that the polygenist theory had been correct in its assertion that man emerged independently from a "widely distributed proto-human ancestry" and that the Caucasian had traversed the various primitive culture stages long before the progenitors of Indian and Negro rose out of bestiality.[33] The trend of progress, aided partly by the extinction of lower races and partly by increasing blood mixture, pointed to a time in the future when the convergent trend would culminate in one human blood and culture.

McGee took a philosophic approach to race struggle and frowned on those weak and trembling of the superior races who shrank from the "self-conjured ghost of imperialism." He felt that the problem of humanity's inferior races was "the strong man's burden." If those who discussed imperialism only knew the much larger ramifications of the topic, they would see it as the spread of a strong man's humanism "enslav[ing] the world for the support of humanity and the increase of human intelligence." Ignoring the law of human progress, "seen through the coordination of other sciences in the Science of Man," such individuals who used the terminology of imperialism, either for or against, failed to understand the process involved. "Imperialism" was a parochial term whose value came only from comprehending the total situation. In the orderly development of all peoples in their vital stages from savagery to barbarism and then to civilization and enlightenment, "imperialism" became as vague as the term "manifest destiny"—an inexplicable ghost caught in the paradox of peoples "rising from

[32] *Ibid.*, 428.
[33] *Ibid.*, 445–46.

plane to plane with a certainty of ultimate union on the highest of the series."

> The white-skinned man indeed leads the world today; but he is not the only burden-bearer. In savagery the strong man leads his fellows, while the weaker fall; in barbarism the strong man leads his family, turning perchance in pity to the weakling; in civilization the strong man supports subjects and feeds their families, and reaches out in helpfulness toward other subjects; but in enlightenment the strong man not only carries the weak until cured or coaxed into strength, but seeks ever to lift to his own plane the world's weaklings, whether white, or yellow, red or black.[34]

The enlightened races, McGee wrote, had more to do than make dutiful subjects of their inferiors; hence, their responsibilities extended far beyond the narrow strictures of the term "imperialism." The "self-taxed task" of the enlightened races, knowing the laws of progress and casting a long shadow from their vantage point in the future, was to "lift the darker fellows to liberty's plane as rapidly as the duller eyes can be trained to bear the stronger light." It was a strong man's burden rather than a white man's burden, a strong man extirpating the bad elements from duller natures and making "leaders of minds in American [Frederick] Douglass and Booker Washington." Understood in these terms, McGee argued that few men today would willingly decry imperialism's purpose and its relationship to progress.[35]

In a speech made in 1903 at the Louisiana Purchase Exposition in St. Louis, McGee reflected upon America's race problems. He surmised that in no other country did the "laggards and leaders comingle so freely." For that very reason, "differences are emphasized and kept in mind." Differences ran to the very fiber of physical and cultural stages in man, which meant that advanced nations attempting to absorb a quota of aliens into their society or trying to maintain a backward race showed the greatest contrast. While endeavoring to lift such peoples to the level of their own culture, a regeneration that extended to both body and mind, work

[34] W J McGee, "National Growth and National Character," address before the National Geographic Society, Mar. 28, 1899, Library of Congress, MSS, fol. 27.
[35] *Ibid.*

and thought, the superior races would discover their standards rising so rapidly that the lower races would "find it hard to keep up." This meant that, despite help, inferior races were "the mental and moral beggars of the community who may not be trusted on horseback but only in the rear seat of the wagon."[36] McGee urged that statecraft and anthropology join hands in the study of human types in an effort to trace the capacity of diverse peoples for progress. Both, standing firmly "on the rock of experimental knowledge," could well define the vigorous and laggard peoples, those "out of harmony with the institutions," and those who would fall behind lawmakers "in such wise that their institutions are inferior to those of progressive nations."[37]

McGee's optimism for humanity's ultimate culture blending stumbled on the question of blood blending. He concluded that intertribal and international blood mixture was beneficial both physically and culturally but that interracial blending was "often apparently injurious, generally of doubtful effect, [and] only rarely of unquestionable benefit." He thought that Frederick Douglass, Booker Washington, Senator Blanche K. Bruce from Mississippi, and poet Paul Laurence Dunbar were fine specimens of white and Negro mixture, but preferred to see them as atypical. Most interracial meetings were illicit "between the lower specimens of one or both lines of blood, so that the evil of miscegenation may well have been intensified." This evil, he warned, forced him to question the intentions of those "eminent Othellos and dignified Desdemonas" who lived as objectionable refugees "domiciled in our national capital."[38]

Geologist and occasional philosopher John Wesley Powell (1834–1902), director of the Bureau of Ethnology and contributor to the *American Anthropologist,* sought to define the exact stage in human progress. An incipient revisionist of Spencerianism, he denied both the Malthusian formula and survival of the fittest in their application to human relationships. Man in the higher civiliza-

[36] W J McGee, "Anthropology and Its Larger Problems," speech made at the Louisiana Purchase Exposition, St. Louis, 1903, Library of Congress, MSS, fol. 27.

[37] W J McGee, "Anthropology and Its Larger Problems," *Science,* XXI (May, 1905), 779–80.

[38] McGee, "Trend of Human Progress," 419.

tions did not compete in the brutal struggle for existence.[39] Having emancipated himself from the cruel workings of nature's indifferent laws, man secured a rank outside and above biotic evolution. Mainly through the acquisition of humanities, the human race removed itself from the "tribes of beasts."[40] Those vestiges of brute competition that still remained in human life existed in criminal behavior, "and to prevent this struggle for existence penal codes are enacted, prisons are built and gallows are erected."[41] Powell, building upon the writings of Lewis Henry Morgan, divided the stages of man's culture into savagery, barbarism, and civilization. The stages of evolution represented "the aggregate of human activities," not "characteristics of individuals."[42] Thus individuals might fail and exhibit retrogression, but culture and races rarely if ever retrogressed; rather, there was a general progress of races and cultures.

Powell felt that much of the previous evidence of retrogression of cultures and races sprang from the misconception that when civilization met with savage or barbaric cultures, it caused the latter's decay and ultimate extinction. Powell felt that the decay of old institutions merely witnessed progression for the savage out of his barbaric culture. Decay did not entail regression for the culture, but, rather, the absorption of wiser opinions, newer institutions, and higher activities. "In all cases," he wrote, "activities borrowed from a higher by a lower culture result in progress" even though individuals within the lower culture may succumb or fall out of the trend of progress.[43]

His conclusions pointed to a wholly new approach in ethnological study. Travelers and scholars in previous ethnological investigations mistook the "jargon of corrupted words" that developed when savage or barbaric cultures met with civilization as having

[39] John Wesley Powell, "Competition as a Factor in Human Evolution," *American Anthropologist*, o.s., I (Oct., 1888), 302, 304; "Professor John W. Powell," *Popular Science Monthly*, XX (Jan., 1882), 390–97; Powell, "Relation of Primitive Peoples to Environment," Smithsonian Institution, *Annual Report for 1895*, 637.

[40] Powell, "Barbarian to Civilization," 98; Powell, "Sociology," 731.

[41] Powell, "Competition," 302.

[42] Powell, "Barbarian to Civilization," 98.

[43] *Ibid.*, 101–2.

the status of language. They inferred from such jargon that tribal languages were not only unstable but "incapable of expressing any great body of thought" necessary to enable change from generation to generation or from one culture scale to another. Powell's attacks centered principally around the conclusions of Spencer, who, he felt, denied "the efficacy of human endeavor."[44] "Man does not compete with plants and animals for existence, for he emancipates himself from that struggle by the invention of arts; and again, man does not compete with his fellow-man for existence, for he emancipates himself from the brutal struggle by the invention of institutions. Animal evolution arises out of the struggle for existence; human evolution arises out of the endeavor to secure happiness; it is a conscious effort for improvement in condition."[45]

Despite Powell's hostility to those who equated human with animal evolution, which called for the survival of the fittest in a struggle for existence, he nevertheless held to mental and bodily improvements through "exercise in the invention of arts, institutions, linguistics, and opinions."[46] He accepted biological improvement as a first step for cultural advancement. Although human evolution resulted in "grades of men" that were essentially intellectual rather than physical, Powell could hardly have disagreed with Spencer's *Principles of Psychology* except perhaps in degree.[47] For Powell, as with Spencer, evolution began as a physical process and through struggle became increasingly intellectual. Powell would have had the greater part of man's evolution reflect an intellectual development that was purposive and distinct from the purely biological realm of tooth and claw. To be sure, he saw cultural evolution as distinct from animal evolution and attacked those who were "overwhelmed with the grandeur and truth of biotic evolution." He believed that culture removed man from comparison

[44] *Ibid.*, 103–4.
[45] Powell, "Competition," 311.
[46] *Ibid.*, 315.
[47] John Wesley Powell, "Relation of Primitive Peoples to Environment, Illustrated by American Examples," Smithsonian Institution, *Annual Report for 1895*, 625; Swan M. Burnett, "The Modern Apotheosis of Nature," *American Anthropologist*, o.s., V (July, 1892), 247–62.

with the beast. Culture was human and "not the development of man as an animal." As a product of human endeavor, it prevented man from becoming the residue of nature's ferment.[48]

Powell was an important bridge between the ideas of the classical evolutionist Morgan and the incipient anticultural evolutionist school of Franz Boas.[49] His foundations were set deep in Darwin's *Descent of Man* and the nineteenth-century belief that physical anthropology, archeology, linguistics, and cultural anthropology were interrelated. At any given level of human evolution there was a "characteristic physical development, a state of material arts, a level of language achievement, and a stage of social organization." The transition from the lowest stage to the highest marked the "unfolding of successively higher levels of intelligence."[50]

Although Powell derived most of his biological concepts from both Spencer and Lester Ward, his ideas of society and the theory of progress came mainly from the influence of Lewis Henry Morgan's *Ancient Society*.[51] The full title of Morgan's book, *Ancient Society, or Researches in the Lines of Human Progress from Savagery through Barbarism to Civilization*, indicates clearly his concept of growth. Due to the psychic unity of all the races, the history of mankind became "one in source, one in experience, and one in progress."[52]

As we re-ascend along the several lines of progress toward the primitive ages of mankind, and eliminate one after the other, in the order in which they appeared, inventions and discoveries on the one hand, and institutions on the other, we are enabled to perceive that the former stand to each other in progressive, and the latter in unfolding relations. While

48 John Wesley Powell, address before the American Association for the Advancement of Science, published in its *Proceedings*, XXXIX (1889), 4–5.

49 Pañchānana Mitra, *A History of American Anthropology* (Calcutta, 1933), 128–29; W J McGee, "Powell as an Anthropologist," Washington Academy of Sciences, *Proceedings*, V (July, 1903), 121–22.

50 Robert E. L. Faris, "Evolution and American Sociology," in Stow Persons, ed., *Evolutionary Thought in America* (New Haven, Conn., 1950), 164–65.

51 John Wesley Powell, "Sketch of Lewis Henry Morgan," *Popular Science Monthly*, XVIII (Nov., 1880), 114–21.

52 Morgan, *Ancient Society*, v–vi, 8.

the former class have had a connection, more or less direct, the latter have been developed from a few primary germs of thought. Modern institutions plant their roots in the period of barbarism, into which their germs were transmitted from the previous period of savagery. They have had a lineal descent through the ages, with the streams of the blood, as well as a logical development.[53]

Oftentimes called "the Tylor of American ethnology," Lewis Morgan (1818–1881) began a career in law and acquired a taste for ethnology as a member of the secret society called the Gordian Knot, organized along the lines of the Iroquois Confederacy. His close relationship with Seneca Indian Ely Samuel Parker and a genuine interest in Indian affairs led him to focus upon the American Indian and the conditions of his civilization.[54] Morgan's inquiries combed the depths of the primitive world in general, and in *Systems of Consanguinity and Affinity of the Human Family* (1871) he affirmed the passage in unilinear evolution through stages of savagery, barbarism, and civilization.[55]

The evolution of man, wrote Morgan, evidenced a physiological development of both his mental and moral powers, and was characterized externally by means of institutions and technology "which express the growth of certain ideas, passions and aspirations."[56] As inventions and discoveries accompanied the development of institutions, so "the human mind necessarily grew and expanded." There was a "gradual enlargement of the brain itself, particularly of the cerebral portion."[57] Growth was slow. The transition from literally nothing to the simplest invention was a monumental epoch in the history of man. Every new minute piece

[53] *Ibid.*, 4.

[54] Arthur C. Parker, *The Life of General Ely S. Parker, Last Grand Sachem of the Iroquois and General Grant's Military Secretary* (Buffalo, 1919), 80–81; J. H. McIlvaine, *The Life and Works of Lewis Henry Morgan* (Rochester, N.Y., 1882), 9; Carl Resek, *Lewis Henry Morgan, American Scholar* (Chicago, 1960), 16, 37–38, 82–83.

[55] Lewis Henry Morgan, *Systems of Consanguinity and Affinity of the Human Family* (Washington, D.C., 1871), 274–75, 504–5; Howard Becker and Harry E. Barnes, *Social Thought from Lore to Science*, 2 vols. (Washington, D.C., 1952), I, 749.

[56] Morgan, *Ancient Society*, 4.

[57] *Ibid.*, 37.

of knowledge added to the complexity of knowledge attained. Progress became a geometric ratio—slow and imperceptible at first but achieving great strides later on.[58]

Those savage societies which were still in the low stage of development would surely make advancements in time but, unfortunately, they would be forever outdistanced by the progress of more advanced peoples. Furthermore, due to their contact with the Aryan and Semitic races, their ethnic arts, languages, and institutions would be destroyed, thus cutting them off from self-development and making them sterile contributors to their own race life.[59] Only the Aryan and Semitic families of man attained the highest point of civilization "through unassisted self-development." The Aryan was the "central stream of human progress" since he alone "provided its intrinsic superiority by gradually assuming the control of the earth."[60]

Morgan, a leader in organizing the anthropological section of the American Association for the Advancement of Science, insisted in *Ancient Society* that the present savage as well as the cerebral ancestor of man were both physically and mentally inferior to the advanced races. From "the great sequence of inventions" which crushed the ignorance that had spread across the "entire pathway of human progress," Morgan sketched human evolution from barbarous to civilized life. He expressed the growth of ideas and society through the study of domestic institutions, in particular, the family. He traced the ancestral experiences of the Aryan nations, though remote from their present advanced status, to similar elements "still preserved in those of savage and barbarous tribes."[61] Human experience, he argued, followed "uniform channels." By virtue of the "specific identity of the brain in all the races of man-

[58] *Ibid.*, 38; Leslie A. White, "Lewis Henry Morgan: Pioneer in the Theory of Social Evolution," in Harry E. Barnes, ed., *An Introduction to the History of Sociology* (Chicago, 1948), 140–41.

[59] Morgan, *Ancient Society*, viii, 371.

[60] *Ibid.*, 4, 7; William H. Holmes, "Biographical Memoir of Lewis Henry Morgan," National Academy of Sciences, *Biographical Memoirs*, VI (1909), 219–39; Powell, "Sketch of Lewis Henry Morgan," 114–20; Bernhard J. Stern, *Lewis Henry Morgan, Social Evolutionist* (Chicago, 1931), 60–61.

[61] Morgan, *Ancient Society*, 8.

kind," mental operations proved uniform when acting in similar conditions.[62]

> Consequently, the Aryan nations will find the type of the condition of their remote ancestors, when in savagery, in that of the Australian and Polynesians; when in the Lower Status of barbarism in that of the partially Village Indians of America; and when in the Middle Status in that of the Village Indians, with which their own experience in the Upper Status directly connects. So essentially identical are the arts, institutions and mode of life in the same status upon all the continents, that the archaic form of the principal domestic institutions of the Greeks and Romans must even now be sought in the corresponding institutions of the American aborigines. . . . This fact forms a part of the accumulating evidence tending to show that the principal institutions of mankind have been developed from a few primary germs of thought; and that the course and manner of their development was predetermined, as well as restricted within narrow limits of divergence, by the natural logic of the human mind and the necessary limitations of its powers.[63]

It was for the good of mankind that the evolutionist and, in particular, the ethnographer sought to expose and interpret the remains "of crude old culture which have passed into harmful superstition."[64] Morgan took pleasure in reciting the atavisms in civilized society. Every human tree had its living fibers rooted deep in history. From the remoteness of its savage era came the tonics and barks which supported its present magnitude. Every people that had risen to eminence drew its nourishment and vigor from a primitive source. But though Morgan extolled the virtues of these hearthstones of a race, he also intended to put a girdling ditch around those remnants that proved harmful. He felt that the Mormons' use of polygamy imparted a relic of ancient savagery "not yet eradicated from the human brain" and harmful to the full development of American life. He explained the circumstances surrounding Mormon society as "outcrops of barbarism . . . explainable

[62] *Ibid.*, 17–18.
[63] *Ibid.*, 553.
[64] Edward B. Tylor, *Primitive Culture* (New York, 1871), 453.

as a species of mental atavism."[65] By pointing out such destructive atavisms Morgan, like later sociologists, sought to transform the science of culture into what Edward Tylor called "a reformer's science."[66]

Another of the leading Jeremiahs of late nineteenth-century ethnology and a forceful contributor to America's concepts of race inferiority was Daniel Garrison Brinton (1837–1899). Born of Quaker descent in Thornbury, Pennsylvania, Brinton served as an assistant surgeon for the federal army from 1862 to 1865, as assistant editor of the *Medical and Surgical Reporter* in 1867, and as its editor in 1874. In 1884 he took the position of professor of ethnology and archeology at the Academy of Natural Sciences in Philadelphia and in 1886 became professor of American linguistics and archeology at the University of Pennsylvania. A prolific writer, his *Notes on the Florida Peninsula* (1859), *Library of Aboriginal American Literature* (1882), and a paper on the mound-builders in the 1886 *Historical Magazine* are but a few of his more noted works. In *The American Race* (1891) he organized the first classification of American aboriginal languages. Throughout the 1890's he contributed voluminous anthropological data to the monthly magazine *Science*.[67]

Brinton emphasized the environmental effect of climate on race progress. Just as the Negro found it hopeless to struggle in climates above the fortieth parallel, so English colonists saw the need to send their children back to Britain for fear of loss of stamina and initiative in an alien tropical climate.[68] The universal struggle for existence had produced paramount differences among the races of man. Some were fortunate in producing that consummate product of creativity, stamina, and superiority while others, weakened by the struggle—the phantom of a remote past—bore the attributes of an unsustained defeat and the marks of inferiority. It was

[65] Morgan, *Ancient Society*, 61.

[66] Tylor, *Primitive Culture*, 453.

[67] An excellent account of Brinton's work in linguistics is found in Regna Darnell, "Daniel Garrison Brinton: An Intellectual Biography" (M.A. thesis, University of Pennsylvania, 1967).

[68] Daniel G. Brinton, *Races and Peoples: Lectures on the Science of Ethnology* (Philadelphia, 1890), 279–80; Brinton, *Negroes* (Philadelphia, 1891), 6.

entirely correct, argued Brinton, to speak of the higher and lower races or progressive and unprogressive races. The American Indian, for example, stood higher than the Australian, Polynesian, and African but lower than the Asian in the race hierarchy. Believing that an intimate relationship existed between the physical and psychical life of man, Brinton suggested the following principal traits as signs of race inferiority, traits which were strikingly similar to those developed by the paleontologist Edward Drinker Cope:[69]

Simplicity and early union of the cranial sutures
Presence of the frontal process of the temporal bone
Wide nasal aperture, with synostosis of the nasal bones
Prominence of the jaws
Recession of the chin
Early appearance, size, and permanence of "wisdom" teeth
Unusual length of the humerus
Perforation of the humerus
Continuation of the "heart" line across the hand
Obliquity (narrowness) of the pelvis
Deficiency of the calf of the leg
Flattening of the tibia
Elongation of the heel (os calcis)

The above traits were "reversions or perpetuations of the apelike (simian pithecoid) features of the lower animals which [were] man's immediate ancestor." Characteristics bearing affinity to the anthropoid apes that existed among the living races became the criteria for judging race gradation as well as aptitude for progress. The arms of the Indian were longer than those of the European but shorter than those of the African. This difference, Brinton argued, was an anatomical evidence of inferiority. "Measured by these criteria," he surmised, "the European or

[69] Brinton, *Races and Peoples*, 40, 42, 47–48; Darnell, "Daniel Garrison Brinton," 117–21; Brinton, *The Basis of Social Relations* (New York, 1902), 49–50. For the attitudes of Cope, see his *Origin of the Fittest: Essays on Evolution* (New York, 1887); "The African in America," *Open Court*, IV (July, 1890), 2400; "The Return of the Negroes to Africa," *ibid.*, III (Feb., 1890), 2110; "Two Perils of the Indo-European," *ibid.* (Jan., 1890), 2052.

white race stands at the head of the list, the African at its foot."[70]

Again, Brinton chose eight particularly basic features of the human form, five of which centered around the head, and set out to explain the gradation from superior to inferior forms.[71]

SKULL	Dolichocephalic, long
	Mesocephalic, medium
	Brachycephalic, broad
NOSE	Leptorhine, narrow
	Mesorhine, medium
	Platyrhine, flat or broad
EYES	Megaseme, round
	Mesoseme, medium
	Microseme, narrow
JAWS	Orthognathic, straight or vertical
	Mesognathic, medium
	Prognathic, projecting
FACE	Chamaeprosopic, low or broad
	Mesoprosopic, medium
	Laptoprosopic, narrow or high
PELVIS	Platypellic, broad
	Mesopellic, medium
	Laptopellic, narrow
COLOR	Leucochroic, white
	Xanthochroic, yellow
	Erythrochroic, reddish
	Melanochroic, black or dark
HAIR	Euthycomic, straight
	Euplocomic, wavy
	Ericocomic, woolly
	Lophocomic, bushy

The races of man, depending upon their stage of culture or somatic growth, progressed at varying rates—from simple arithmetical progression of the savage, to geometrical progression of the half-cultured, to saltatory progression (permutation) of the enlightened races. Arithmetical progression had little influence in

[70] Brinton, *Races and Peoples*, 47–48.
[71] *Ibid.*, 49–50.

the general culture of a people. The smelting of iron, for example, long known among the African tribes, had little benefit for the race beyond its "immediate convenience for weapons." Likewise the Chinese, who had known of the compass and gunpowder long before the European, had been unable to utilize them as potential civilization builders.[72] Neither the American Indian nor the Negro built "empires" of significant duration. "The limitations of the racial mind," wrote Brinton, "were such that a complex social organization was impossible for them."

> In the forms of their highest governments, those of the Aztecs, Mayas, and Peruvians, we see repeated on a large scale the simple and insufficient models of rude hunting tribes of the plains. This is also true of the black race of Africa. The powerful monarchies which at times have been erected in that continent over the dead bodies of myriads of victims have lasted but a generation or two. . . . Indeed the law of "thus far shalt thou go and no farther" tells the story of most of the failures of races and peoples. They fell through mental inability to succeed. They had reached the natural limit of their activities.[73]

In a short pamphlet entitled *Negroes,* published in 1891, Brinton compared the physiology of Caucasian and Negro for signs of inferiority. Interestingly enough, he conceded that the Negro was not susceptible to acute alcoholism as was the Caucasian. Nonetheless, even this criterion for physiological differences did not act to the Negro's favor; rather, Brinton argued, the Negro's seeming immunity was attributable to his general physiological inferiority and, in particular, to "the inferior susceptibility of [his] nervous system." Taking issue with the autopsy findings of Civil War physician Sanford B. Hunt, Brinton judged that the Negro had smaller lungs and a larger liver but that there was no appreciable difference in brain weight.[74] According to Brinton, brain weight was of little consequence in the Negro's mental makeup. The Negro was "pacific and cheerful" and, although anxious for education, was "unwilling to make the necessary mental effort to obtain

[72] Brinton, *Basis of Social Relations,* 79–80.
[73] *Ibid.,* 71.
[74] Brinton, *Negroes,* 6–7.

it."[75] Like so many other anthropologists of the nineteenth century, Brinton added that the Negro child was precocious and, in fact, on an equal footing with white children of his own age, but that his progress usually ended at puberty. After puberty "there supervenes a visible ascendency of the appetites and emotions over the intellect, and an increasing indisposition to mental labor."[76]

It was this physiological and mental change at puberty that made Brinton all the more adamant that there could be no mixture of the races in the United States. Miscegenation brought an "indelible degradation" to the descendants of the white partner in such a marriage.[77] He did not mean that miscegenation should be prohibited absolutely. He meant his warning only for the United States and its superior Caucasians. In areas of Brazil, Peru, and Mexico mixtures of African with American races had often improved race stocks. The Cafusos of Brazil, the Zambos of Paraguay, and the Chinas of Peru were "finely formed and vigorous" and had "repeatedly taken precedence in political and social life over the pure descendants of the European colonists."[78]

Like Darwin before him, Brinton saw that the cross between Caucasian and Negro produced a hybrid wholly deficient for the American climate. Like Darwin too, he took note of the deficient physical vigor of mulattoes on the Gold Coast, unable to bear up in the peculiar geographic region.[79] Brinton found that the American mulatto, susceptible to consumption and scrofula, lived for a shorter time than either the full Negro or Caucasian.[80] He wrote, "It is essential, also, to remember that it is the inferior race only which reaps the psychical advantage. Compared to the parent of the higher race, the children are a deteriorated product. Only

[75] *Ibid.*, 8.

[76] *Ibid.*, 11; Brinton, *Basis of Social Relations*, 136.

[77] Brinton, *Races and Peoples*, 287.

[78] *Ibid.*, 284–85.

[79] Charles Darwin, *The Descent of Man* (London, 1871), 171.

[80] Brinton, *Races and Peoples*, 283–84. Brinton's analysis mirrored the findings of the Civil War studies. See Sanford B. Hunt, "The Negro as a Soldier," *Anthropological Review*, VII (Jan., 1869), 40–54; Benjamin A. Gould, *Investigations in the Military and Anthropological Statistics of American Soldiers* (New York, 1869); J. H. Baxter, *Statistics, Medical and Anthropological, of the Provost Marshal–General's Bureau*, 2 vols. (Washington, D.C., 1875).

when contrasted with the average of the lower race can they be expected to take some precedence. The mixture, if general and continued through generations, will infallibly entail a lower grade of power in the descent. The net balance of the two accounts will show a loss when compared with the result of unions among the higher race alone."[81]

To preserve the wholesomeness of the American stock from the folly of hybridization, Brinton looked to the American woman. "It is to the woman alone of the highest race that we must look to preserve the purity of the type, and with it the claims of the race to be the highest."[82] Curiously, his call was but a reflection of the South's appeal through the antebellum years and Reconstruction for the preservation and purity of the southern white woman. The gyneolatry that Brinton defended from the position of science was a conventional notion mouthed by aristocrat and poor white alike. Science, however, gave to the Caucasian the symbols of pure nationality and helped raise the cult of white womanhood to the scientific idolatry of race.[83]

Nineteenth-century ethnologists paid obsequious attendance to mind growth and physical characteristics of races in their study of the stages of human progress from savagery to enlightenment. They saw the minds of particular culture grades responding in similar manner to like stimuli. The mental development of the races of the world were finely distinct and easily recognizable from both physical characteristics and human response patterns. The generalizations of their quarter-century's observations on mankind saw the human mind as a product and mirror of ancient burdens and present labors. Indeed, every mind was of open-ended design and of accidental value, not anticipated in the future, not concluded in the past. Culture grades furnished man not only with a certain intelligence but also with peculiar physiological features, stereotyped reactions, and wonderfully matching institutions. Weak and nerveless races plowed similar human treads. Language,

[81] Brinton, *Basis of Social Relations*, 155.
[82] Brinton, *Races and Peoples*, 287.
[83] Wilbur J. Cash, *The Mind of the South* (New York, 1941), 87–88, 117–19, 131, 309–11, 339–40, 347.

music, and human wants reflected a nation's achievement as they did the unseen baggage of some ancient gesture or unconscious act long past. Events which made annals of nations were but shadows of human moments, both present and past, reflections of human experiences formed in the pale of universal laws and intersecting history at their highest moments.

V *From Biology to Sociology: Spencer and His Disciples*

TWO OF THE MOST recognized popularizers of evolution in the late nineteenth century, Herbert Spencer (1820–1903) and his American disciple John Fiske (1842–1901), argued that man's cultural life developed according to the same evolutionary laws applicable in the physical world. With the comparative method as their tool, they bridged the physical and social sciences, maintaining that many indications of parallel cultural development in both living and extinct societies suggested that cultures underwent tensions and struggles similar to those in the biological world. That widely separated peoples developed identical patterns of culture could not be explained wholly on the basis of diffusion from a single center of origin.[1] The social Darwinist or Spencerian of

[1] Joseph LeConte, "Scientific Relation of Sociology to Biology," *Popular Science Monthly*, XIV (Feb., 1879), 427; Herbert Spencer, "The Relation of Biology, Psychology, and Sociology," *ibid.*, L (Dec., 1896), 163–72; Anna Tolman Smith, "A Study of Race Psychology," *ibid.* (Jan., 1897), 354–61; Charles A. Ellwood, *Some Prolegomena to Social Psychology* (Chicago, 1901), 36–37.

the era, viewing evolution as an all-encompassing formula, made no distinction between social and organic development. In looking at the savage, for example, as "an arrested or rather retarded stage of social development," the social Darwinist made an easy transition into both psychology and biology by seeing a correlation in "the evolution of the human mind that an examination of the embryo supplies of the evolution of the human body."[2] Evidence in one sphere of human activity became a common denominator for analogous combinations in sister sciences. Like the later Freudian psychoanalysts who argued "for innate action-patterns or specific human instincts as underlying cultural activity," so the social Darwinist "linked up the stages of cultural evolution with corresponding stages in psychical or mental development."[3] Cultural organization became strictly analogous with physiological organization, and facts uncovered in one were readily conceded as evidence in the other. "The extension of the doctrine of evolution to psychical phenomena," wrote Fiske, "was what made it [evolution] a universal doctrine."[4]

FROM BIOLOGY TO PSYCHOLOGY

According to Spencer, biology involved primarily the internal phenomena of living things while psychology served as the connecting link between internal phenomena and the environment. Although biology also described phenomena in the environment (since the life of every organism involved an adaptation to exterior environmental actions), Spencer maintained that psychology was something beyond this simple relationship. Psychology did not belong exclusively to the objective world nor to the subjective world but, taking a term from each, occupied itself with the correlation

[2] James G. Frazer, *The Scope of Social Anthropology* (London, 1908), 7.

[3] David Bidney, "Human Nature and Cultural Process," *American Anthropologist*, n.s., XLIX (July-Sept., 1947), 395; Morris E. Opler, "Cultural and Organic Conceptions in Contemporary World History," *ibid.*, XLVI (Oct.-Dec., 1944), 448; Franz Boas, "The Methods of Ethnology," *ibid.*, XXII (Oct.-Dec., 1920), 311.

[4] John Fiske, *A Century of Science*, vol. X of *John Fiske's Miscellaneous Writings*, 12 vols. (Boston, 1902), 47.

of the two.[5] Psychology was "not the connexion between the internal phenomena, nor is it the connexion between the external phenomena; but it is *the connexion between these two connexions.*"[6] Though Spencer spoke of psychology as a distinct science, he argued that there was no sharp line separating biology from psychology, "only different groups of phenomena broadly contrasted but shading off into one another." Having accepted evolution, he saw no break or change "from one group of concrete phenomena to another," only "a universal process, one and continuous throughout all forms of existence."[7] Biology and psychology were distinguishable only as "specialized parts of the total science" taking account of the continuous transformation of the universe.[8]

Unlike simple organic structures in which the relationship between inner changes in adaptation to outer phenomena was purely mechanical (plant life relating to the cycle of the seasons), those creatures with higher intelligence showed a capacity to correspond to the time cycles of nature in a nonmechanical manner. The more developed the intellectual faculty, the higher the order of correspondence of organisms to time. Among the savage races of man, for example, methods for estimating epochs corresponded to the coincidence of bird migration, flooding, or plant life. Each longer time sequence implied not only a higher grade of civilization but a further adjustment of internal relations to newer external relations. It meant "additional series of vital actions" in the brain and therefore "an increased number of heterogeneity of the combined changes which constitute life."[9] The Hottentots, a step removed from the more primitive Australians, were capable of making correspondence to the temporal processes of the environment from both astronomical and terrestrial phenomena.[10] From the nomadic root diggers and insect eaters to the semicivilized races which built

[5] Herbert Spencer, *The Principles of Psychology*, 2 vols. (New York, 1870–1872), I, 130.

[6] *Ibid.*, 132.

[7] *Ibid.*, 136.

[8] *Ibid.*, 137.

[9] *Ibid.*, 327.

[10] *Ibid.*, 325.

huts, accumulated property, and stored commodities for future periods, there was a corresponding difference in time recognition. Similarly, the seventeenth-century astronomer who computed the trajectory of a comet which took centuries to elapse between prediction and fulfillment evidenced a further specialization of correspondence from one whose environment was almost homogeneous in space and time to one "whose environments are homogeneous in Space but heterogeneous in Time."[11]

Extending his argument for specialization a step further, Spencer suggested that it would certainly not aid the savage to know the seasons of the year or the tide changes unless he was similarly able to put such knowledge to use. If he had not the dexterity to cast fish hooks, the perception of the hook would be meaningless. In other words, the extension of correspondence in space and time marked similar and partly reciprocal changes in physiological specialization. Race development, if and when it occurred, progressed through slow but successively higher physiological and psychological stages. Differentiation of perception had to allow for subsequent differentiation of actions, an ability to modify conduct to correspond to new conditions. "The more various the impressions receivable from surrounding things," wrote Spencer, "the greater must be the number of modifications in the stimuli given to the motor faculties; and hence, the greater must be the tendency towards modified actions in the motor faculties." Progress in one involved progress in the other, "in respect both of activity and complexity."[12]

There were instances, wrote Spencer, where minds of inferior peoples were incapable of responding to complex relations. Sandwich Islanders could "learn by rote with wonderful rapidity, but will not exercise their thinking faculties"; the Australians had no power of concentration or integration of separate ideas; and Negro children in the United States educated along with white children did not "correspondingly advance in learning—their intellects be-

[11] *Ibid.*, 329; Herbert Spencer, *Principles of Sociology*, 3 vols. (New York, 1876–1897), I, 84–85; Spencer, *Descriptive Sociology*, 11 vols. (London, 1873–1925), III, 43.

[12] Spencer, *Principles of Psychology*, I, 354–55; Spencer, *Principles of Sociology*, I, 86.

ing apparently incapable of being cultured beyond a particular point."[13] The small-brained savage without even a vocabulary including words meaning "justice," "sin," or "mercy" could hardly account for human actions that manifested intellectual or moral complexity. The manifestations of complexity exhibited in the large-brained races, on the contrary, constituted a long developmental process. To suggest, therefore, that an inferior small-brained Hottentot could be transplanted at birth to a civilized community and develop on an equal status with its adopted society presupposed that knowledge resulted strictly from experience. Such an inference, Spencer warned, ignored the mental evolution that accompanied "the autogenous development of the nervous system."

Doubtless, experiences received by the individual furnish the concrete materials for all thought. Doubtless, the organized and semi-organized arrangements existing among the cerebral nerves, can give no knowledge until there has been a presentation of the external relations to which they correspond. And doubtless, the child's daily observations and reasonings aid the formation of those involved nervous connexions that are in process of spontaneous evolution; just as its daily gambols aid the development of its limbs. But saying this is quite a different thing from saying that its intelligence is wholly produced by its experiences.[14]

The human brain was the combined register of past race evolution and present experiences. The race struggles of the distant past transmitted a brain mass whose qualities improved through the frequency of added experiences. Strengthened or weakened from use or disuse, the brain was then bequeathed to a future generation. In this manner the European inherited some thirty cubic inches of brain more than the lowly Papuan, a situation significant enough to account for the Newtons and Shakespeares in one and the inability of the other "to count up to the number of their fingers."[15] As life advanced from savage to civilized man, advanc-

[13] Spencer, *Principles of Psychology*, I, 368; Spencer, *Principles of Sociology*, I, 59, 93–94, 101.

[14] Spencer, *Principles of Psychology*, I, 469–70.

[15] *Ibid.*, 471; Spencer, *The Study of Sociology* (New York, 1874), 34–35.

ing civilization offered "more numerous experiences to each man, as well as accumulations of other men's experiences, past and present." Experiences became more heterogeneous "as by degrees civilization supplies them and develops the facilities for appreciating them."[16] The widening of experiences produced more varied associations of ideas, which diminished "rigidity of belief" so evident in the savage and also allowed for a "plasticity of thought" to accompany increasing brain activity.[17] "The ever-multiplying connexions of ideas that result," Spencer wrote, "imply ever-multiplying possibilities of thought."[18] Such continual processes in civilization made "beliefs more modifiable—so furthering other changes, mental and social."[19]

The tendency of weak, reflex-action races to mimic the motions of the more developed types indicated another proof of inferiority among the races of man. Spencer suggested the degrees of decrease of "this irrational mimicry" as a possible scale on which the stages of social organization could be drawn. He also believed that race analysis by means of philology would offer a similar graph of cultural progression. It was possible to draw a scale of mental development by studying the degrees of generality and abstraction in language vocabulary. Then, too, the animal kingdom yielded an enormous wealth of information from which parallel associations could be drawn which explained analogous differences among the races of man. Inferior animals had a mental type which was almost entirely guided by reflex actions. Only in a small degree were they capable of changing the mode of their experiences. As the nervous structure developed in higher animals their actions were less confined to pre-established limits and "individual experiences take longer and longer shares in moulding the conduct." There was an ability in more complex mental types to profit from past experiences.[20] The races of man, argued Spencer, paralleled this animal

[16] Spencer, *Principles of Psychology*, II, 524–25.
[17] *Ibid.*, 536.
[18] *Ibid.*, 525.
[19] *Ibid.*, 536.
[20] Herbert Spencer, "The Comparative Psychology of Man," *Popular Science Monthly*, VIII (Jan., 1876), 260–61, 266; Spencer, *Principles of Sociology*, I, 93–94.

development. The semicivilized nations, "characterized by a greater rigidity of custom," were less capable of modifying their ideas and habits to present or future experiences. Marked by an early precocity and arrested mental development at puberty, they soon relaxed into a relatively automatic nature, incapable of responding to stimuli in other than a reflex-response pattern. Just as an infant showed small persistence in any one thing (wanting an object and then abandoning it for something new), so the inferior races exhibited resistance to "permanent modification." Lacking intellectual persistence, "they [could not] keep the attention fixed beyond a few minutes of anything requiring thought even of a simple kind." Intensity of any sort produced exhaustion.[21]

Spencer's scheme of evolution tolled a note of pessimism for the less civilized peoples of mankind. Progress in intellectual and social development depended upon a natural movement through successive stages from homogeneity to heterogeneity. A savage was unable to live with civilized man as an equal, since civilization's complex associations could not be comprehended by his inferior brain whose capacity was geared to a far simpler framework of association. This also meant that, for all practical purposes, evolution concerned only the Caucasian. Whatever progress might come to the savage could accrue only in insignificant stages which in no way approached the accelerated state of the Caucasian's evolution. Larger brain weight in the Caucasian permitted an increase in experiences and representativeness of thought on a scale far greater than that possible in the savage. While the Caucasian advanced in an almost geometric progression, the savage's slow ascent in the scale of unilinear evolution made him entirely insignificant in the race struggle with the nations of large-brained peoples. True, the savage provided a unique opportunity for the Caucasian to study his own race history, since all races had ascended the same unilinear scale. But beyond the savage's ethnographic and historical significance in explaining the

21 Spencer, "Comparative Psychology of Man," 260–61; Spencer, "The Comparative Psychology of Man," *Mind*, I (Jan., 1876), 11.

Caucasian's remote past, the small-brained peoples were of little importance, perhaps none at all.[22]

The world applauded Spencer's *Synthetic Philosophy* as the most momentous intellectual conquest of the day. "Probably no philosopher," wrote editor Henry Holt, "ever had such a vogue as Spencer had from 1870 to 1890."[23] Journals ranging from *Biblioteca Sacra* to *Macmillan, Nation, Popular Science Monthly,* and *Contemporary* published or explained sections of his architectonic system. He published his *First Principles* in 1863, followed by *Classification of the Sciences* (1864), *Principles of Biology* (1864–1867), *Principles of Psychology* (1870–1872), *The Study of Sociology* (1874), *Principles of Sociology* (1876–1897), and *Principles of Ethics* (1892–1893). Spencer enjoyed a greater acceptability in the United States than in England, and by the end of the century over 300,000 volumes of his works had been sold in America.[24]

The full title of Spencer's *Descriptive Sociology: Encyclopaedia of Social Facts Representing the Constitution of Every Type and Grade of Human Society, Past and Present, Stationary and Progressive, Classified and Tabulated for Easy Comparison and Convenient Study of the Relations of Social Phenomena,* speaks more for the author's cultural evolutionism than any discursive explanation could. Arranged in three divisions (uncivilized societies, extinct or decayed civilized societies, and recent or still flourishing civilized societies), Spencer's sociology was essentially a compendium of attributes and descriptions taken from travelers and catalogued according to physical, emotional, and educational characteristics, political structure, rites, habits, superstitions, habitations, weapons, implements, and esthetic products. These descriptions of the races were as varied as they were contradictory. In one instance the Fuegians were described as a "timid race"

[22] Spencer, *Principles of Sociology,* I, 109; John C. Greene, *The Death of Adam: Evolution and Its Impact on Western Thought* (Ames, Iowa, 1959), chap. 10.

[23] Henry Holt, *Garrulities of an Octogenarian Editor* (Boston, 1923), 298.

[24] George Sarton, "Herbert Spencer," *Isis,* III (1921), 375–91; J. W. Burrow, *Evolution and Society: A Study in Victorian Social Theory* (Cambridge, 1966), chap. VI; Thomas A. Goudge, "Philosophical Trends in Nineteenth Century America," University of Toronto, *Quarterly,* XVI (1946–1947), 141.

always speaking "in a whisper" but in another instance as "loud and furious talkers." Frequent emphasis was given to the anthropoidal characteristics of races, particularly the use of the feet for such work as cutting flesh (Dyaks), drawing a bow (Veddahs), and picking things up from the ground (Veddahs).[25] There was also much emphasis placed upon the facial angle, prognathism, precocity of children, ability to imitate, sense of smell, and concepts of time.[26]

Spencer borrowed heavily from the writings of earlier anthropologists and travelers, adopting their observations in part or as a whole to illustrate the pervasiveness of biological law. He borrowed from Heinrich Lichtenstein (1780–1857) and Sir John Barrow (1764–1848) on the Bushmen, William Lewis Herndon (1813–1857) on the Brazilians, Robert Southey (1774–1843) on the Tupis, Martin Dobritzhofer (1717–1791) on the Abipones, William Gifford Palgrave (1826–1888) and Sir Richard Francis Burton (1821–1890) on the Bedouins and prairie Indians, Sir Francis Galton (1822–1911) on the Damaras, and Peter Kolb (1675–1726) on the Hottentots. He was also deeply indebted to Henry Rowe Schoolcraft for information concerning the Iroquois and other American Indian tribes. Spencer had nothing but travelers' narratives upon which to base his knowledge of primitive cultures, and in his enthusiasm to find evolution working everywhere he did not subject his borrowed evidence to rigorous analysis. Since Spencer lacked direct knowledge of these cultures, *Principles of Psychology* as well as *Principles of Sociology* bore the traces of secondhand speculation. Gilbert Malcolm Sproat's account of the Ahts of North America was a case in point. Spencer became intensely interested in Sproat's conclusion about the savage. "The native mind," wrote Sproat, "seems generally asleep." Aroused to conversation, the native wearied quickly, "particularly if questions are asked that require efforts of thought or memory on his part."[27] Similarly, Spencer quoted from *Travels in Brazil, 1817–1820* by Johann Baptist von Spix and Karl Friedrich

[25] Spencer, *Descriptive Sociology*, III, 1, 3.
[26] *Ibid.*, VI, 1.
[27] Sproat quoted in Spencer, *Principles of Sociology*, I, 94–95.

Philip von Martius on the Brazilian Indian: "Scarcely has one begun to question him about his language, when he grows impatient, complains of headache, and shows that he is unable to bear the exertion."[28] These conclusions as well as countless others became the funding source for Spencer's *Synthetic Philosophy*, a compendium of testimonies drawn together to form a calculated portrait of race character. Spencer's use of materials was so sweeping in application that critics challenged his concepts more by rhetorical debate than by piecemeal verification, a situation which was common to the whole structure of anthropology in its early years.[29]

The solution to race inferiority, Spencer felt, was not in mixing the diverse races. Union of widely divergent varieties was "physically injurious" to the offspring, producing a "worthless type of mind—a mind fitted neither for the kind of life led by the higher of the two races, nor for that led by the lower—a mind out of adjustment to all conditions of life." On the other hand, mixtures of slightly divergent types were physically beneficial to the life of a race, producing mental types having "superiorities" in adjusting to new conditions in life. Spencer mentioned Samuel Smiles's *The Huguenots* (1873) and Sir Francis Galton's *English Men of Science* (1874) as proof of the good effects of mixture of slightly divergent varieties of the same stock.[30] Edward Youmans, one of

[28] Spix and Martius quoted in *ibid.*, 95. Spencer also borrowed heavily from Sir Francis Galton, *The Narrative of an Explorer in Tropical South Africa* (London, 1853); Sir Richard Francis Burton, *Selected Papers on Anthropology, Travel and Exploration* (London, 1924); Burton, *Personal Narrative of a Pilgrimage to El-Medinah and Meccah*, 3 vols. (London, 1855–1856); William G. Palgrave, *Narrative of a Year's Journey through Central and Eastern Arabia*, 2 vols. (London, 1865); Martin Dobritzhofer, *An Account of the Abipones, and Equestrian People of Paraguay* (London, 1822); Robert Southey, *History of Brazil*, 3 vols. (London, 1817–1822); William Lewis Herndon, *Explorations of the Valley of the Amazon*, 2 vols. (Washington, D.C., 1853–1854); Peter Kolb, *The Present State of the Cape of Good-Hope*, 2 vols. (London, 1731); Heinrich Lichtenstein, *Reisen im sudlichen Africa, in den Jahren 1803, 1804, 1805 und 1806*, 2 vols. (Berlin, 1811–1812); Sir John Barrow, *Travels into the Interior of Southern Africa . . .* (London, 1806).

[29] Richard H. Shryock, discussion of paper by John C. Greene in Marshall Clagett, ed., *Critical Problems in the History of Science* (Madison, Wis., 1959), 451.

[30] Spencer, "Comparative Psychology of Man," *Popular Science Monthly*, 262–63; Spencer, "The Development of Political Institutions," *ibid.*, XVIII (Jan., 1881), 292.

Spencer's many propagandists, clarified the Englishman's remarks for his American audience by warning that the mixture of northern European stock with Asian, African, or inferior European stocks was "extremely injurious" to the Aryan race in the United States. Such "is a corollary from biological facts."[31]

The generation that read Darwin could hardly avoid reading the works of Spencer. Yet Charles Darwin could never convince himself that the theory Spencer developed had much validity. Personally, he found him a particularly egotistical man.[32] As a scientist Darwin felt "aggrieved" that Spencer would palm off so much deductive speculation as irrefutable biological law. "If he had trained himself to observe more," wrote Darwin to J. D. Hooker, "even at the expense . . . of some loss of thinking power, he would have been a wonderful man."[33] Beatrice Webb, a lifelong friend of Spencer, was similarly critical of his work. She found his explanations "immensely impressive" to the "enthusiastic novice in scientific reasoning," but could not help believing that he had spent a lifetime "engaged in the art of casuistry."[34]

> Partly in order to gain his approbation and partly out of sheer curiosity about the working of his mind, I started out to discover, and where observation failed, to invent, illustrations of such scraps of theory as I understood. What I learnt from this game with his intellect was not, it is needless to remark, how to observe—for he was the most gullible of mortals and never scrutinized the accuracy of my tales—but whether the sample facts I brought him came within the "law" he wished to illustrate. It was indeed the training required for an English lawyer dealing with cases, rather than that of a scientific worker seeking to

[31] Edward Youmans, *Herbert Spencer on the Americans and the Americans on Herbert Spencer* (New York, 1883), 20; Herbert Spencer, *An Autobiography*, 2 vols. (New York, 1904), II, 61–62, 111–12, 115–17; Max Fisch, "Evolution in American Philosophy," *Philosophical Review*, LVI (1947), 358–61; Grant Overton, *Portrait of a Publisher* (New York, 1925), 50; John Fiske, *Edward Livingston Youmans* (New York, 1894), 115; Charles M. Haar, "Edward L. Youmans: A Chapter in the Diffusion of Science in America," *Journal of the History of Ideas*, IX (1948), 193–213.

[32] Charles Darwin, *The Autobiography of Charles Darwin* (London, 1958), 108.

[33] Darwin to J. D. Hooker, Dec. 10, 1886, in Francis Darwin, *The Life and Letters of Charles Darwin*, 2 vols. (New York, 1888), II, 239.

[34] Beatrice Webb, *My Apprenticeship* (London, 1926), 25.

discover and describe new forms of life. What he taught me to discern was not the truth, but the relevance of facts. . . .[35]

Both Edward B. Tylor and Thomas Henry Huxley were critical of Spencer's synthetic handling of evidence. Tylor, who himself had enriched unilinear evolution with rules of organic correlation, felt that Spencer tended to ignore all evidence that did not conform to the exact stages of intellectual evolvement set down by his system.[36] Similarly, Huxley argued that Spencer concocted his theory from his "inner consciousness" rather than from empirical evidence. "He is the most original of thinkers," wrote Huxley, "though he has never invented a new thought." Spencer was a "great constructor," bringing together older ideas to act as the "component factors" in his new synthetic system.[37] Josiah Royce, one of Spencer's biographers, called him "a philosopher of a beautiful logical naivete." Though he presented a theoretically unified and orderly exposition of his system, which he "always had at perfect control," it was nevertheless not "the same as the perfection of one's theory."[38] Ironically, however, criticism of Spencer had little bearing on his concepts of race inferiority and much less upon the derivation of his racial ideas. The subject of race inferiority was beyond critical reach in the late nineteenth century.

John Fiske became an enthusiastic convert to Spencer during his college days at Harvard. One of the earliest advocates of evolution, he continually amazed the Cambridge community with his appetite for explaining history, religion, science, language, philosophy, and anything else that interested him under the formula of cosmic law. Like Spencer, he was a compiler of secondhand information and was forever anxious to carry the history of the Anglo-Saxon peoples to higher and higher plateaus of veneration. Basically a popularizer rather than an original thinker, he borrowed heavily from Spencer, Edward A. Freeman, Sir Henry Maine, Lewis Henry Morgan, Alfred R. Wallace, Thomas H. Huxley, Wil-

[35] *Ibid.*, 26–27.

[36] Edward B. Tylor, "Mr. Spencer's 'Principles of Sociology,'" *Mind*, II (Apr., 1877), 144.

[37] Webb, *My Apprenticeship*, 27.

[38] Josiah Royce, *Herbert Spencer: An Estimate and Review* (New York, 1904), 116.

liam Stubbs, and a host of others. His career in public lecturing re-echoed the writings of Freeman on the Aryan sources of American institutions, the Teutonic prejudices of James K. Hosmer, the environmentalist theories of Harvard geologist Nathaniel S. Shaler, and the pastoral loves of James Russell Lowell and Francis Parkman. President of the Immigration Restriction League in the 1890's, Fiske gave his blessings to the champions of Brahmin chauvinism seeking to perpetuate an enduring image of Anglo-Saxon birthright in America. His writings as well as his countless lectures appeared intellectually profound to the superficially educated American middle class seeking a foothold in the transitional years of evolutionary thought. His audiences left his lectures satisfied but, like Fiske himself, unwilling to look beneath the tranquil surface for fear of discovering the uncharted map of a chance universe. Like Spencer's synthetic philosophy, Fiske's cosmic philosophy generalized to the point of encompassing all and nothing; yet he pleased his audience with a consoling picture of perplexing intellectual ferment. He had so well assimilated the views of Spencer that he had a surprising ability to draw out implications just being formulated by Spencer himself.[39]

Like Spencer, Fiske began his explanations of racial differences with the brain. The Teuton's 114 cubic inches of brain, compared with the Australian's 70 cubic inches, indicated, aside from mass, a structural complexity that left the Australian closer to the chimpanzee, with 35 cubic inches.[40] Borrowing many of his ideas from Lyell's *Antiquity of Man,* Fiske compared the non-Aryan Hindu with an Englishman and remarked that "the difference in volume of brain between the highest and lowest man is at least six times as great as the difference between the lowest man and the highest ape."[41] In his study of brain capacity Fiske was fond of

[39] Milton Berman, *John Fiske: The Evolution of a Popularizer* (Cambridge, Mass., 1961), chap. II, 94–95, 210–11, 268; Russel B. Nye, "John Fiske and His Cosmic Philosophy," Michigan Academy of Science, Arts, and Letters, *Papers,* XXVIII (1942), 685–98; Holt, *Garrulities of an Octogenarian Editor,* chap. XIX; Barbara M. Solomon, *Ancestors and Immigrants: A Changing New England Tradition* (Cambridge, Mass., 1956), 61–64, 68–69, 104.

[40] John Fiske, *Outlines of Cosmic Philosophy,* vols. I-IV of *John Fiske's Miscellaneous Writings,* IV, 48–49.

[41] *Ibid.,* 93.

quoting Galton's *Tropical South Africa.* Along with Lester Ward, Charles Darwin, Spencer, and others, he drew his analogy of brain complexity from the famous comparison of the primitive Damara seeking a solution to a simple mathematical problem and the dilemma of Galton's dog Dinah.

> Once while I was watching a Damara floundering hopelessly in a calculation on one side of me, I observed Dinah, my spaniel, equally embarrassed on the other. She was overlooking half a dozen of her new-born puppies, which had been removed two or three times from her, and her anxiety was excessive, as she tried to find out if they were all present, or if any were still missing. She kept puzzling and running her eyes over them, backwards and forwards, but could not satisfy herself. She evidently had a vague notion of brain. Taking the two as they stood, dog and Damara, the comparison reflected no great honour on the man.[42]

From the incident Fiske concluded that the capacity for progress was not common among all of mankind. The smaller-brained races were "almost wholly incapable of progress, even under the guidance of higher races." "The most that can be said for them," he wrote, was "that they are somewhat more teachable than any brute animals."[43] Man's integration into complex social organizations, the mark of civilization, removed the individual from his static, habitual existence and made him associate directly in more complex experiences and in ideas of far-reaching variety. The "decomposition and recombination of thoughts" in this new encounter facilitated abstraction and generalization and also the "plasticity of thought." In this widening of human experiences there existed the genesis of progress and the conspicuous "chasm which divides man intellectually from the brute."[44]

Infancy became a critical period in race development. Like society moving from simple to complex associations and aggregations, infancy was an interval in which "the nerve connections and

[42] *Ibid.*, 50, quoting Galton, *Tropical South Africa*, 132.

[43] Fiske, *Outlines of Cosmic Philosophy*, IV, 53; John H. Van Evrie, *White Supremacy and Negro Subordination* (New York, 1868), 219–20; Fiske, *The Discovery of America, with Some Account of Ancient America and the Spanish Conquest*, 2 vols. (New York, 1892), I, 100; II, 212.

[44] Fiske, *Outlines of Cosmic Philosophy*, IV, 91.

correlative ideal associations for self-maintenance are becoming permanently established."[45] But such capacity for nerve development depended on a prior intelligence or capability—brain mass. Accordingly, Fiske doubted the Lockean supposition of a *tabula rasa.* The infant's mind was not a blank sheet but, rather, "a sheet already written over here and there with invisible ink, which tends to show itself as the chemistry of experience supplies the requisite conditions." In other words, the mind of the infant was "correlated with the functions of a more complex mass of nerve tissue which already has certain nutritive tendencies."[46] Nor did Fiske subscribe to the ideas of Leibniz and Kant concerning intuitional knowledge. Ideas were not innate but the result of "nutritive tendencies in the cerebral tissue, which have been strengthened by the uniform experience of countless generations." After juxtaposing Locke and Kant in the nature-nurture problem, Fiske then tried to bring Locke's learning by experience and Kant's innate ideas into a higher synthesis through the doctrine of evolution. "In learning," he wrote, "we are merely acquiring latent capacities of reproducing ideas; and . . . beneath these capacities lie more or less nutritive tendencies which are transmissible from parent to child."[47]

In organisms with little or no infancy (codfish, turtle, flycatcher) the nervous system developed before birth, a situation which maintained a conservative tendency in heredity and allowed little possibility for modification after birth. Life was "predetermined by the careers of . . . ancestors," and there was only a narrow area for environmental circumstances to modify it.[48] But in those higher life forms with extended infancy nerve connections and associations formed after birth, and heredity yielded to the more immediate impressions made by circumstances. The longer the infancy, the more was the opportunity for modification by environment and the greater was the possibility for variation from ancestral forms. Long infancy not only increased the possibility

[45] *Ibid.*, 131.
[46] *Ibid.*, III, 236–37.
[47] *Ibid.*
[48] Fiske, *Excursions of an Evolutionist,* vol. VII of *John Fiske's Miscellaneous Writings*, 283–84.

for a more "plastic" mental type but it also encouraged inventiveness and individuality, the backbone of race progress.[49] By means of the comparative method and the accounts of nineteenth-century travelers, Fiske carried his theory of infancy into an analysis of the so-called inferior races.

For example, Alfred Wallace wrote of finding a "little half-nigger baby" when he was out shooting in the Malay Archipelago.[50] Actually he had killed an orang and had taken its baby to care for. He noted how the baby's characteristics were almost human and how it even demanded a pillow when it slept. "I am sure nobody ever had such a dear little duck of a darling of a little brown hairy baby before," he wrote.[51] From such observations Fiske, and others reading his account, felt that there was not much difference between the orang and the lowest type of man. The helpless "little half-nigger baby" that Wallace mothered for several months seemed to Fiske an adequate explanation of the gradation from the ape-man to the Aryan.[52] Exemplifying a period of limited plasticity, the infant orang was unable to feed itself or walk without aid even at three months. Born without developed prenatal capacities, the nervous system had only potentialities which had "to be roused according to his own individual experience."[53] But the ultimate potentiality of the orang was limited by its brain mass, the product of centuries of prehistoric struggle and variation.

In proceeding from the apes to man, not only the structure and capacity of the brain changed but the period of infancy lengthened. While the lowest races in the human family had longer infancy than the anthropoids, it was still shorter than the Caucasian's infancy. Furthermore, intellectual development in the lower races appeared to stop at the time of puberty—a time marked by

[49] Fiske, *Darwinism and Other Essays*, vol. VIII of *John Fiske's Miscellaneous Writings*, 41; Berman, *John Fiske*, 98; Spencer, "Comparative Psychology of Man," *Mind*, 9; Howard Becker and Harry E. Barnes, *Social Thought from Lore to Science*, 2 vols. (Washington, D.C., 1952), I, 736–37.

[50] Alfred Wallace, *My Life: A Record of Events and Opinions*, 2 vols. (New York, 1905), I, 343; Fiske, *Excursions of an Evolutionist*, 285–86.

[51] Wallace, *My Life*, I, 345.

[52] Fiske, *Excursions of an Evolutionist*, 285–86.

[53] Fiske, *A Century of Science*, 104.

the closing of the cranial sutures. According to Fiske, whose opinions in this area were borrowed from the writings of Spencer, Robert Dunn, Frederick W. Farrar, and Filippo Manetta, the length of infancy in the various races corresponded to the complexity of brain convolutions and creases, the furrows increasing in depth as one approached the higher races.[54] Fiske, like Spencer, fell back on craniometry and the anatomical differences in brain weight and convolutions in his explanation of why certain inferior races were unable, despite evolution, to achieve greatness. Spencer believed that there was a direct relationship between the "extra mental mass" of the higher races and their significant superiority in energy and mental vigor.[55] For Fiske, the cerebral development accounted for "all the conspicuous physical peculiarities of man except his bare skin." The increase of intelligence had a direct bearing upon facial angle, size of the jaw, and inclination of the forehead. Intelligence came first and disuse of primitive bodily features followed. Man's intellectual conquest of nature occasioned new forms of implements, different foods, and the diminution of organs no longer needed in primitive circumstances. "It is probable," Fiske argued, "that increased frontal development had directly tended, by correlation of growth, to diminish the size of the jaws, as well as to push forward the bridge of the nose."[56]

Progression from inferior to superior stock corresponded directly to increased size and capacity of the cerebrum over the cerebellum.

> Continuing to grow by the addition of concentric layers at the surface, the cerebrum becomes somewhat larger in birds and in the lower animals. It gradually covers up the optic lobes, and extends backwards as we pass to higher mammalian forms, until in the anthropoid apes and in man it covers the whole upper surface of the cerebellum. In these highest animals it begins also to extend forwards. In the chimpanzee and gorilla the anterior portion of the cerebrum is larger than in inferior mammals; but in these animals, as in the lowest races of man, the frontal extension is but slight, and the forehead is both low and narrow.

[54] Fiske, *Outlines of Cosmic Philosophy*, III, 96.

[55] *Ibid.*, IV, 99; Spencer, "Comparative Psychology of Man," *Popular Science Monthly*, 258–59; Spencer, *Principles of Sociology*, I, 56–57.

[56] Fiske, *Outlines of Cosmic Philosophy*, IV, 99.

In civilized man, the anterior portion of the cerebrum is greatly extended both vertically and laterally. As already observed, the most prominent physiological feature of human progress has been the growth of the cerebrum. The cranial capacity of the average European exceeds that of the Australians and Bushmen by nearly forty cubic inches; and the expansion is chiefly in the upper and anterior portions.[57]

Over all the earth's surface, Fiske saw that the thousands of years of prehistoric struggle had determined the capacity of the various races of man. The African, Polynesian, and American races were in an arrested state, far beneath that of the civilization of the higher races. Even continual contact with the more advanced Caucasian could not remove them from savagery except by direct intervention. Most of those races which succeeded in going beyond the condition of savagery had, he added, "been arrested in an immobile type of civilization, as in China, in ancient Egypt, and in the East generally." Only the Aryan and some Semitic, Hungarian, and Finnic tribes had shown any "persistent tendency to progress."[58]

Through the widespread popularization of Fiske the conclusions of somatology were woven into concepts of psychic evolution to show an almost overwhelming presumption that evolutionary science had found laws fixing some races, such as the Negro, outside the possibility of the Caucasian's progress and civilization. Both Spencer and Fiske saw themselves as representatives of "the highest cultural development towards which all other more primitive cultural types tend," which meant the imposition of an orthogenetic structure tending toward the rationalization of western European civilization.[59]

FROM PSYCHOLOGY TO SOCIOLOGY

Fiske described sociology as "an extension of the principles of biology and psychology to the complex phenomena furnished by

[57] *Ibid.*, III, 194–95.
[58] *Ibid.*, IV, 3, 4.
[59] Boas, "Methods of Ethnology," 312.

the mutual reactions of intelligent organisms upon each other."[60] In joining sociology by means of the comparative method to its sister sciences of biology and psychology, he offered a soothing palliative to American concepts of racial inferiority in the late nineteenth century. While psychology had become handmaiden to biology, endeavoring "to interpret the genesis of intellectual faculties and emotional feelings in the race, and their slow modification through countless generations," sociology, on the other hand, studied the phenomena produced by the reactions of individuals upon each other and then generalized and established laws for the understanding of the phenomena.[61] This meant that the task of the sociologist, in the widest sense of the term, aimed "at discovering the general laws which have regulated human history in the past, and which, if nature is really uniform, may be expected to regulate in the future."[62] Through greater understanding of the conditions under which certain phenomena occurred, the sociologist hoped, by means of regulatory actions, "to make . . . volitions count for something in modifying them."[63]

Grounded in biological and psychological science, the Spencerian sociologists saw man as a cell in a social organism. By endeavoring to understand the individual's biological and psychological makeup, they believed that they could assign "to each individual that special function to which he is best adapted" and thus direct "the path which society must follow to effect reforms."[64] Relying

[60] Fiske, *Outlines of Cosmic Philosophy*, II, 41; Gustave LeBon, "The Influence of Race in History," *Popular Science Monthly*, XXXV (Aug., 1889), 495–96; Frazer, *Scope of Social Anthropology*, 6.

[61] Fiske, *Outlines of Cosmic Philosophy*, II, 53; Ellwood, *Some Prolegomena to Social Psychology*, 36–37; G. Archdall Reid, "The Biological Foundations of Sociology," *American Journal of Sociology*, XI (Jan., 1906), 532–54; Carlos C. Closson, "A Critic of Anthropo-Sociology," *Journal of Political Economy*, VIII (June, 1900), 397–99.

[62] Frazer, *Scope of Social Anthropology*, 4; Roscoe Hinkle, *The Development of Modern Sociology* (New York, 1954), 4–9.

[63] Fiske, *Outlines of Cosmic Philosophy*, III, 250–51; John Wesley Powell, "Sociology, or the Science of Institutions," *American Anthropologist*, n.s., I (July, 1899), 475–77.

[64] Achille Lora, "Social-Anthropology—a Review," *American Anthropologist*, n.s., I (Apr., 1899), 284, 288; Frederick J. Teggart, "Prolegomena to History: The Relation of History to Literature, Philosophy, and Science," University of California, *Publications in History*, IV (1916), 244–45.

on the comparative method, the Spencerians did not hesitate to use historical societies as well as organic life to explain and predict human relationships. And under the guise of finding general laws of nature operating in both the social and biological organism, they were really engaged in an effort to prescribe what America ought to be—an idea that incorporated the subjective prejudices and assumptions about themselves and others, and a belief that science was an objective tool engaged in an "uninvolved" diagnosis of empirical reality.

Late nineteenth-century sociologists like Gerrit Lansing, Charles Ellwood, Sarah E. Simons, and F. W. Blackmar offered an analysis of the races in America that bridged the gulf between the Negro and the Caucasian. Borrowing from John Wesley Powell's stages of growth from barbarism to civilization, they generally felt that the Indian had in only a few cases come out of savagery into the stage of barbarism. Most Indians were crude specimens of prehistoric time held under the hammer of an overbearing environment and succumbing to race suicide at the touch of the Caucasian's civilization. The Indian, like the Negro, held a place in the gradation of the races of man. Sociologists like Blackmar of the University of Kansas and Simons of Washington, D.C., reflected the thought of the earlier American ethnologist Henry R. Schoolcraft, who remarked that the Indian and the colonist represented the "alpha and omega of the ethnological chain."[65] Like entomologist Augustus Radcliffe Grote (1841–1903), many sociologists saw the Indian as a child who had "not passed the mental state of our own children" and who, afraid of change, stood steadfast in a suicidal conservatism.[66] Yet sociological studies of the American Indian lacked many of the stereotypes characteristic of the scientific investigations of the Negro, the focus of most nineteenth-century racial thinking. Because physiological differences were

[65] Henry R. Schoolcraft, *The American Indians: Their History, Condition and Prospects, from Original Notes and Manuscripts* (Rochester, N.Y., 1851), 369; J. Henry Gest, "Our Indian Mythology," *Popular Science Monthly*, XXIII (Aug., 1883), 528.

[66] A. R. Grote, "The Early Man of North America," *Popular Science Monthly*, X (Mar., 1877), 582–83; John Fiske, *The Discovery of America*, 2 vols. (New York, 1892), I, 100; II, 212.

seemingly more visible between Negro and Caucasian, scientists found it far simpler to presume race inferiority from the Negro's anthropometric statistics and then infer the existence of a scale of races between him and the Caucasian.[67] Given the structure of nineteenth-century anthropometry, scientists found it easier to argue Negro inferiority from facial angle, prognathism, and brain weight than to argue inferiority of Indian, Chinese, and Malayan using the same devices. Craniometric differences were not as marked in statistical studies of the "average man" of such stocks as Chinese, Malayan, and Indian. The physical anthropologists, therefore, generally shunned direct contrast.

Spencerians offered an answer to the craniometric dilemma of the physical anthropologists by working backward through the scientific series. Taking generalizations derived from contemporary social behavior, they inferred both biological and psychological causation for race differences among Chinese, Indian, and other nonwhite groups. Differences, however slight from anthropometric measurement, became measurably significant from their studies of race behavior. Recognizing, as had Spencer and Fiske, the artificial boundaries in the division of the sciences, they offered to untangle the phenomena of race inferiority by elaborating on the discoveries in their own discipline. The derivative science of sociology ascertained facts in the realm of contemporary social patterns that offered firm corroboration of what had been only suggested by evolutionist studies in other sciences. Sociologists felt that if their evidence could be incorporated with the conclusions of the sister sciences, scientists as well as social scientists could disentangle the relationships of frequently obscure racial phenomena. In building the science of sociology, sociologists believed that their field could use evidence only partially revealed in the other sciences.[68]

In their efforts to understand the conditions which best explained Caucasian race progress, sociological thinkers like George

[67] Daniel Brinton, *The American Race: A Linguistic Classification and Ethnographic Description of the Native Tribes of North and South America* (New York, 1891), 42.

[68] Hinkle, *Development of Modern Sociology*, 8–9.

S. Painter of the New York State College for Teachers, Paul S. Reinsch of the University of Wisconsin, F. W. Blackmar, and Fernand Papillion adopted the notion of self-development as the most productive feature of race evolution. In analyzing the reasons for the Caucasian's greater advancement, they inquired into the differences between self-education of the Caucasian and the manner in which he in turn educated the intermediate race stocks. Of the three processes of race education—self-development, imitation, and compulsory activity—only the Caucasian had accomplished anything by the first method. The Caucasian, wrote Blackmar, was the lone "man in evolution"; his present race level unfolded as an unbroken thread linking its present generations with ancestors of centuries past. All other races, having failed in the primary struggle at some prehistoric point in time, looked to advancement by means of imitation or compulsory activity. For the Indian, imitation was superficial and of value only as his blood had "become mixed with that of the white race." The Caucasian's use of compulsory activity in Indian education, forcing the Indians "out of their natural gait," had produced only incidental success. Begun and forcibly carried out by the Caucasian, compulsory education was an extreme humanitarian measure designed to save the Indian from self-destruction. Hoping "to arouse the [Indian's] latent energies of his own nature," wrote Blackmar, the Caucasian sought to bring him to the level of self-development by drawing him through the lower stages of advancement.[69]

As for the exhibition of imitativeness by the lower races—a term first used by Walter Bagehot in *Physics and Politics* (1872), later by Gabriel Tarde in *Les lois de l'imitation* (1890 and 1895) and James Mark Baldwin in *Mental Development in the Child and the Race* (1895)—sociologists disdained any inherent worth in the phenomenon.[70] It was doubly easy, wrote Sarah Simons, for the

[69] F. W. Blackmar, "Indian Education," American Academy of Political and Social Science, *Annals*, I (May, 1892), 813–14; Blackmar, *The Elements of Sociology* (London, 1905), 231–32, 237–38; Charles A. Ellwood, "Review of William Benjamin Smith's The Color Line: A Brief in Behalf of the Unborn," *American Journal of Sociology*, XI (Jan., 1906), 570–75.

[70] Franklin H. Giddings, "Modern Sociology," *International Quarterly*, II (1900), 550.

Negro to imitate the culture of the Caucasian, since he possessed "no transmitted culture of his own to act as a barrier to the adoption of a new life."[71] His adaptability, imitativeness, and patience made it easy for him to absorb new habits and customs, but the assimilative ability was only superficial because of his inferior mental capacity and primitive instincts. The same analogy was true for the Indian. Praising the teleological designs in natural selection, Spencerians affirmed the value of adjustment solely through long periods of evolution. Any method of progress outside a purely laissez-faire evolutionary framework, not incorporating a struggle for existence, provoked their deepest skepticism. "Every imitation," wrote W. M. Flinders Petrie, "was a direct injury to character. . . . [Imitation] teaches a man to trust to some one else instead of thinking for himself; it induces a belief in externals constituting our superiority, while foresight and self-restraint are the real roots of it; and it destroys all chance of any real and solid growth of character which can flourish independently. A native should always be discouraged from any imitation, unless he attempts it as an intelligent improvement on his own habits."[72]

A "series of instinctive impulses" limited the process of imitation, wrote Charles Ellwood, professor of sociology at the University of Missouri. Long a disciple of Spencer, he maintained that imitativeness by one race of another occurred only on the surface and took no account of the race "instincts"—organic sympathy, mental attitudes, religious feelings, and morality—that developed with an organism "through a process of evolution by natural selection." Development by imitation was obviously restricted by race instincts and tendencies, Ellwood argued. Otherwise "we should expect children of different races, when reared in the same cultural environment, to develop the same general mental and moral characteristics." Such was not the case, nor could it be without

[71] Sarah E. Simons, "Social Assimilation," *American Journal of Sociology*, VII (Jan., 1902), 543; Jabez L. M. Curry, "The Negro Question," *Popular Science Monthly*, LV (June, 1899), 179–80; F. C. Spencer, *Education of the Pueblo Child* (New York, 1899), 88.

[72] W. M. Flinders Petrie, "Race and Civilization," Smithsonian Institution, *Annual Report for 1895*, 598.

contradicting the foundations of the biological and psychical order of development. The Negro child, "even when reared in a white family under the most favorable conditions, fails to take on the mental and moral characteristics of the Caucasian race."[73] While the "mental attitudes" of Negroes and Indians concerning society, persons, and religion became the same as those of the Caucasian by the process of imitation, and despite the fact that their natural instincts might modify in the course of generations, nonetheless voluntary or even enforced imitativeness could not conceal a "race habit of a thousand generations." Though encrusted with the virtues of a superior race, they were but vagabonds profiting from a perverted philanthropy, deprived of the intrinsic value of their outward show of merits. Ellwood felt that the lack of self-development in superficial assimilation explained the "strong tendency to reversion."[74] The revival of voodooism and fetishism among Negroes in the South, despite the veneer of Christian influences, indicated that their instinctive tendencies were far below those of the superior race. In general, imitativeness by an inferior race of its superior left a significant gulf between the advanced society and the inferior race's more primitive instincts. To permit imitation by an inferior race of a higher society without any recognition by the lower race of the "life-process" involved in race achievement was to think that a sound body could restore an unsound mind. To force education upon an inferior race or to allow it to imitate, wrote Ellwood, drove an insensible wall between the forces that molded human society and the "forces which have shaped evolution in the past."[75] Only a wise philanthropy which did not "over-

[73] Charles A. Ellwood, "The Theory of Imitation in Social Psychology," *American Journal of Sociology*, VI (May, 1901), 735. Ellwood depended heavily upon psychologists James Mark Baldwin and C. Lloyd Morgan for his theory of imitation. See Baldwin, *Social and Ethical Interpretations in Mental Development* (New York, 1897), 8, 70, 230, 378; Baldwin, *Mental Development in the Child and the Race* (New York, 1895), chaps. 9 and 12; Morgan, *Habit and Instinct* (London, 1896), chaps. 8 and 15.

[74] Ellwood, "Theory of Imitation," 735–36; Nathaniel S. Shaler, ed., *The United States of America*, 2 vols. (New York, 1894), I, 49; F. W. Chapman, "European and Non-European; or the Relation of the White to the Colored Races," *Education*, XX (Sept., 1899), 1824; Paul B. Barringer, *The American Negro: His Past and Future* (Raleigh, N.C., 1900), 5, 13, 23.

[75] Ellwood, "Theory of Imitation," 740.

look the biological consequences of its activities" would really benefit "degenerate and defective human beings."[76]

A true civilizing process, wrote John Roach Straton of Mercer University in Macon, Georgia, did not come from "artificial development from without, but a gradual and harmonious growth from within."

> Plato's dwellers in the cave could not be suddenly transferred from their accustomed darkness to the dazzling light on the outside. The African cannot be lifted to the plane of the Anglo-Saxon by the use of either logarithms and Greek roots or formulae for cultivating a field or constructing a pair of shoes. The Anglo-Saxon has reached his present high civilization after a long and laborious struggle upward. Through a series of well defined steps, he has risen from barbarism to his present plane. The system in which he now dwells is the teleological outcome of all that has gone before, and consequently the white-man of to-day is thoroughly suited to his environment. Now, it is reasonable to think that, since Anglo-Saxon civilization is thus the culmination of a series of steps, all the steps must be taken before it can safely be reached. To suddenly introduce another race, therefore, to any step in the series, and then to attempt to hurry it over the other steps in the hope of having it reach and occupy the culminating one, must be a hopeless undertaking. The evolutionary process cannot be supplanted by artificial stimulants. Should we wonder, then, that our educational efforts in behalf of the negro seem to have failed of their intended purpose? Nay, more—does not the history of races show that the effort on the part of a superior people to lift up inferiors at a single stroke not only fail but established conditions which lead to the actual destruction of the weaker race.[77]

What began as the white man's burden to lift the inferior races into the age of enlightenment became, in reality, the black man's death.

[76] D. Collin Wells, "Social Darwinism," *American Journal of Sociology*, XII (Mar., 1907), 716; George S. Wilson, "How Shall the American Savage Be Civilized?" *Atlantic Monthly*, L (Nov., 1882), 607. Wilson's argument was meant primarily as a caution toward unwise education of the Caucasian woman. There exists an interesting parallel between the concepts of race inferiority and of the woman's place in nature in the late nineteenth century. See William I. Thomas, "The Mind of Woman and the Lower Races," *American Journal of Sociology*, XII (Jan., 1907), 435–69.

[77] John R. Straton, "Will Education Solve the Race Problem?" *North American Review*, CLXX (June, 1900), 793–94; Elbridge S. Brooks, *The Story of the American Indian: His Origin, Development, Decline and Destiny* (Boston, 1887), 41–42, 281–82.

Failing to develop safeguards to the heightened dangers consequent to civilization, the weaker races fell easily to the temptations, dangers, and strains of civilized communities. Beneath the surface of imitativeness, wrote Straton, loomed the debilitating manifestations of ethical and physical decay, weakened fertility, venereal disease, and dangerously high infant mortality. While there were always certain individuals (usually those strengthened in "mental vigor by infusions of white blood") who exhibited progress under white tutelage, the Negro race, like other inferior races, showed tendencies which gave "color to the fear that they are a decaying people."[78]

Jabez L. M. Curry, chairman of the educational committee of the John F. Slater Fund, a philanthropic organization designed to aid the emancipated Negro, reflected much of the same thinking. True human development, he wrote, came not from external causes but "from voluntary energy, from self-evolved organizations of higher and higher efficiency."[79] Behind the Negro were centuries of barbarism, idolatry, fetishism, and all the consequences of this dangerous heredity. The institution of slavery had prevented natural race growth, and the Negro's subsequent freedom and citizenship, "with natural weaknesses uncorrected, with loose notions of piety and morality and with strong racial peculiarities and proclivities," had left him with the feeble moral sense "common to all primitive races."[80] Before advocating further integration into the government, commerce, and civilization of the white man, one had first to understand the physical, hereditary, and racial characteristics of the Negro.[81] Individual cases (Blanche Bruce, Booker T. Washington, Frederick Douglass) did not demonstrate the permanent capacity of the race or its mental possibilities. The

[78] Straton, "Will Education Solve the Race Problem?" 797–98; Daniel G. Brinton, *The Basis of Social Relations* (New York, 1902), 160–61; A. D. Mayo, "How Shall the Colored Youth of the South Be Educated?" *New England Magazine*, XVII (Oct., 1897), 213.

[79] Jabez L. M. Curry, "Difficulties, Complications, and Limitations Connected with the Education of the Negro," John F. Slater Fund, *Occasional Papers*, no. 5 (1895), 5.

[80] *Ibid.*, 7.

[81] Jabez L. M. Curry, "Education of the Negroes Since 1860," John F. Slater Fund, *Occasional Papers*, no. 3 (1895), 5–7.

general uplifting of the race might require centuries. Since differences presently existing between the African and the European were the product of thousands of years of race development, "centuries may pass before their relations as neighbors and fellow citizens have been duly adjusted."[82]

One of the more significant sociological developments that grew out of Spencer's *Principles of Sociology,* as well as his *Principles of Biology,* concerned the implications of Chinese immigration into the United States. Actually, sociological arguments against the Oriental were but corollaries to similar investigations dealing with the Indian and Negro. Dr. John H. Van Evrie, known for his book *White Supremacy and Negro Subordination,* accepted the possibility that the Chinese had attained a semblance of civilization in the past but concluded that whatever merit existed in their history was due to the presence of Caucasian blood. Confucius as well as other renowned ancestors known to the modern Chinese were really Caucasian, argued Van Evrie, "and what shadowy and uncertain historical data they now possess are therefore likely to have originated from these sources."[83] Physician Albert S. Ashmead of New York, carrying his racial feelings a step further, urged restriction of Orientals on the basis that they were universally subject to the "torpid musings and prurient sensual disquietude" which came with puberty.[84] On the whole, however, the Chinese posed a peculiar dilemma for American race concepts.

Sociologists knew well the aptitude of the Chinese for adaptability in various climates, industry, and intellectual ability. Their facial angle, cephalic index, prognathism, and past history were not without praise.[85] Besides claiming one of the greatest empires, the Chinese had given an emphasis to education in government

[82] Curry quoting Lord Bryce, "Education of the Negro," 11; W. T. Harris, "The Education of the Negro," *Atlantic Monthly,* LXIX (June, 1892), 721–36.

[83] Van Evrie, *White Supremacy and Negro Subordination,* 80.

[84] Albert S. Ashmead, "Relation of Syphilis with Japanese Racial Peculiarities and Customs," *Atlanta Journal-Record of Medicine,* VIII (Sept., 1906), 395; Ashmead, "Notes and News," *American Anthropologist,* o.s., IX (June, 1896), 219.

[85] M. Bond, "Observations on a Chinese Brain," *Brain,* VII (1894), 37; Frederick W. Farrar, "Aptitudes of Races," Ethnological Society of London, *Transactions,* V (1866), 124; "The Chinese Immigration," *The New Englander,* XXIX (Jan., 1870), 3–4; W. T. Brigham, "Measurements of 300 Chinese," Boston Society of Natural History, *Proceedings,* XI (1866), 98–100.

that put the Gilded Age to shame. For obvious reasons, American sociologists avoided these criteria to establish the Oriental's inferiority. Instead they chose a slippery combination of Spencerian struggle of the fittest, where environment had a profound effect on the modification of individuals, and Galton's law of inheritance. While the former provided the dynamic for Anglo-Saxon or Aryan race development, the latter became the tool for rationalizing the Oriental's lack of development in the modern world.[86] What separated Chinese from Anglo-Saxon, argued the sociologist Gerrit Lansing, was his absolute adherence to tradition and customs. While the West gloried in change and progression, the Chinese, by virtue of pointless ancestral worship, exhibited a suicidal tenacity to race characteristics which, like those prominent in the Indian and Negro, fell outside the laws of development, survival of the fittest, and variation. With an element of armchair aloofness, Lansing concluded that the Chinese as a race exhibited an ability to transmit race characteristics that persisted long after they had "passed the condition of usefulness."[87] Their present civilization, customs, and laws reflected a mental character that was charming but inconsequential, irresponsible, and silly. Many of their traits "which were anciently of great benefit, [and] still transmitted by inheritance [became] injurious by interfering with the introduction of new forms of greater utility." While the Negro could hardly assimilate without endangering the physical and mental makeup of the Caucasian, the Chinaman was unwilling to assimilate due to innate racial conservatism. Since the permanence of America's political future depended upon the "homogeneity of its civilization," the Chinese immigrant, by virtue of centuries of continued inbreeding, would stand apart, and by the law of heredity would try to force American institutions, as well as their progressive character, to submit to the "older and more deeply rooted" Chinese conventions

[86] Simons, "Social Assimilation," 539; Richard Mayo-Smith, "Assimilation of Nationalities in the United States," *Political Science Quarterly*, IX (Sept., 1894), 431; Francis Galton, *Hereditary Genius: An Inquiry into Its Laws and Consequences* (London, 1869), 336.

[87] Gerrit L. Lansing, "Chinese Immigration: A Sociological Study," *Popular Science Monthly*, XX (Apr., 1882), 723; Reid, "Biological Foundations of Sociology," 546.

and conservatism. The conclusion of Chinese inferiority, unlike the conclusions derived by physical anthropologists concerning the Negro, was made on the basis of a sociological judgment of contemporary Chinese social behavior. Chinese inferiority resulted from a cultural bondage held fast by portentously invoked ancestral worship and made evident in nearly all their present history.[88]

The Aryans, so went the sociological argument of Lansing, had developed to a progressive state long before they left their ancestral homes in Central Asia. Aryan language "passed out of the monosyllabic stage" as, ever inquiring, these peoples adapted to new environments without holding firm to quadrumanous race characteristics. The Chinese, on the other hand, retained their "simple monosyllabic form" which afforded the strongest reason for classifying them as a primitive race and mere spectators in the survival of the fittest. Their blighted language, according to Lansing, became an external caricature of their biological race development. The preservation of primitive culture traits illustrated the law of heredity: "characters which have been long transmitted are more persistent than those of more recent origin."[89] The Chinese, added sociologist William I. Thomas of the University of Chicago, afforded "a fine example of a people of great natural ability letting their intelligence run to waste from lack of a scientific standpoint."

> [T]hey are not defective in brain weight, and their application to study is long continued and very severe; but their attention is directed to matters which cannot possibly make them wise from the occidental standpoint. . . . Their attention to Chinese history is great, as benefits their reverence for the past, but they do not organize their knowledge, they have no adequate textbooks or apparatus for study, and they make no clear distinction between fact and fiction. In general, they learn only rules and no principles, and rely on memory without the aid of reason, with the result that the man who stops studying often forgets

[88] Lansing, "Chinese Immigration," 724; Spencer, *Principles of Sociology*, I, 78; Percival Lowell, "The Soul of the Far East," *Atlantic Monthly*, LX, (Nov., 1887), 614.

[89] Lansing, "Chinese Immigration," 732.

everything, and the professional student is amazingly ignorant in the line of his own work.[90]

Galton's law of inheritance defined the dilemma of the contemporary Chinaman. In contrast to the Anglo-Saxon the suffocating weight of ancestral worship had burdened the Chinese with a race character that looked almost wholly to the past and gained little from present experiences. Joseph LeConte admired the Chinese for their race development but felt that their extreme rigidity under alien race contact left them unsympathetic and unimpressionable to further race evolvement.[91] This rigidity fixed upon the Chinese who came to America a conservative temperament, unwilling and even unable to change or to assimilate. Assimilation, if it occurred, would be by the superior Aryans, bending and compromising their advanced principles and healthy plasticity to the intransigent soul of the Oriental.[92] By mixing Anglo-Saxon blood with that of the Chinese, wrote physician J. P. Widney, "a race utterly without the instincts of representative government in their mental constitution" would result, a race of people incapable of self-government who would force a worthless hybridization upon the body politic.[93] As far as Lansing was concerned, the Oriental's traditional blind dependence upon authority, reflected in the lack of a "word for liberty in their language," would assuredly eclipse America's struggle to preserve freedom, independence, and individualism for its people. "In the involuntary conflict ensuing, those characters which originated before the dawn of ancient history, and have been strengthened through the inheritance of unnumbered generations, would persist with greater force than those new and changing characters which seem by comparison like the fashions of a season. . . . The new society would assume more the character of its persistent than its more yielding part. Intense

[90] Thomas, "Mind of Woman," 454.

[91] Joseph LeConte, "The Race Problem in the South," Brooklyn Ethical Association, no. 28 (1893), 362.

[92] A. P. Peabody, "The Chinese in San Francisco," *American Naturalist*, IV (Jan., 1871), 663–64; Edward Ross, 'The Causes of Race Superiority," American Academy of Political and Social Science, *Annals*, XVIII (July, 1901), 87–88.

[93] J. P. Widney, "The Chinese Question," *Overland Monthly*, n.s., II (Dec., 1883), 629; Report of the Joint Special Committee to Investigate Chinese Immigration, *Senate Report No.* 689, 44th Cong., 2nd sess. (1876–1877), 586–87.

conservatism would check the progress of reform and improvement."[94]

According to the sociological argument of Lansing, the American constitutional form of government stood as symbol of the people's democratic ideology, representing all sections and all classes. Elements within such an organism which refused assimilation, however, threatened the very foundations of government and its prevailing ideology. The same principles that applied to the evolution of organisms governed the evolution of society. Since a society existing as an aggregate "necessitates the homogeneity of the parts composing it," Lansing argued, the people must "be of the same race and civilization, and in their institutions, laws and customs, represent those instincts and temperaments which are characteristic of their race."[95] It was not enough to allow the chances of natural selection to bring about the solution to the problem of the Mongol. The Chinaman's propensity to save and his willingness to live on next to nothing gave him an unhealthy advantage in the world of the Anglo-Saxon who had come to demand more from life. The Mongol was the thistle in the race life of the Anglo-Saxon. "Shall we pluck it up, as does the wide husbandman," wrote M. J. Dee, "or shall we withdraw the intelligence of artificial selection from the environment, and leave the battle to the chances of natural selection alone?" The life-sustaining faculties of the Anglo-Saxon were far removed from the primitive art of sustaining life exhibited by the Mongol.[96] Just as Rome had fallen because of the admixture of unassimilative elements, so America, despite her boundless optimism and humanitarian feeling for the world's destitute, might suffer the same fate.

The Chinese, like the Negro and Indian, became for many American sociologists a grotesque prehistoric people, breaking into their contemporary world and threatening the foundations of America's democratic experiment. The Chinese represented a stage in human evolution in which stagnation had long since re-

[94] Lansing, "Chinese Immigration," 734.

[95] *Ibid.*, 725; Simons, "Social Assimilation," 797, 799.

[96] M. J. Dee, "Chinese Immigration," *North American Review*, CXXVI (May-June, 1878), 526.

placed development. China as well as the other backward nations had emerged in the nineteenth century as "a fossilized representative of an antique system, physically active but mentally inert."[97] In this manner nineteenth-century Spencerians offered contributions to the century's social-scientific theories regarding the various "inferior" races. Relatively plain in their speech and phraseology, they sought to create a social-scientific foundation for determining the best social structure in America. Since the Caucasian embodied the highest level of adaptability, inventiveness, and democratic principles, American Spencerians sought to limit opportunity to him alone and to restrict from participation those races bound by imitative tradition or prodded by compulsory education. The science of society in the late nineteenth century not only confirmed in a circular fashion the assumptions of biological racial inferiority but also helped to frame the ideology of disfranchisement and immigration restriction.

[97] Charles Morris, *Man and His Ancestor* (New York, 1900), 194.

VI *Portraits of Nineteenth-Century Academic Thinking on Race*

THE RACIAL THINKING of Nathaniel Southgate Shaler (1841–1906), dean of the Lawrence Scientific School at Harvard, of geologist and evolutionary idealist Joseph LeConte (1823–1901) of the University of California, and of paleontologist Edward Drinker Cope of the University of Pennsylvania epitomized in many ways the most "scientifically" accepted attitudes of the late nineteenth century on the Negro, the immigrant, and the so-called "inferior races." Their intellectual rationalizations, *a priori* values, and race analyses mirrored the cumulative mind of the century's intellectual concepts of race. Like John Fiske, Charles Ellwood, Daniel Brinton, W J McGee, and other nineteenth-century scientists and social scientists, these three men rejected the cold mechanism of natural selection and took a decided Lamarckian or neo-Lamarckian approach in their analysis of nature and man's capabilities. In marked similarity to other pre-Mendelian scientists of their generation, they explained man's behavior as the product of and interaction with his social environment. With the aid of a social-scientific methodology, they helped to rationalize the century's belief in the physical and mental diversity of races in order to justify contemporary racial thought.[1]

Lamarck's emphasis upon the direct action of the environment rather than struggle for existence, inheritance of acquired characteristics, cumulative effect of use and disuse of organs, and adaptation of organs to the environment reflected the American emphasis upon manipulative possibilities and the environmentalist tradition in the behavioral sciences in America. In part, LeConte, Shaler, and Cope were reacting against blind, nonpurposive evolution. They saw a broader social significance to human evolution—an evolution in which the inheritance of acquired characteristics affirmed their attachment to rational social behavior. Their admiration of evolution did not end in the abandonment of intellect to fortuitous circumstance but, rather, in a desire to preserve through social action the ideal of the nineteenth-century enlightened man in the polyglot world. As scientists and social scientists they looked to the immense engine of evolution and environmental change for the intricate formulae of social order, reform, and progress.[2]

JOSEPH LECONTE: THE SOCIOLOGY OF RACE

Born in 1823 of French Huguenot descent in Liberty County, Georgia, Joseph LeConte spent a lifetime in science and education, first in the South and later in California, where he lived until his death in 1901. His father, Louis LeConte, had migrated from New Rochelle, New York, to an estate in Woodmanston, Georgia, in 1810. A scientist in his own right and a student of Dr. David

[1] Edward J. Pfeifer, "The Genesis of Neo Lamarckism," *Isis*, LVI (1965), 156–67; William D. Armes, ed., *The Autobiography of Joseph LeConte* (New York, 1903), 150–51; Joseph LeConte, "Origin of Organic Forms: Is It by Natural or Supernatural Process?" *Overland Monthly*, XVIII (Aug., 1891), 198–203.

[2] George W. Stocking, Jr., "Lamarckianism in American Social Science: 1890–1915," *Journal of the History of Ideas*, XXIII (1962), 239–56; Stow Persons, *American Minds: A History of Ideas* (New York, 1958), 242–44; Stocking, *Race, Culture, and Evolution: Essays in the History of Anthropology* (New York, 1968), chap. 10.

Hosack in medicine, Louis LeConte spent his years dabbling in botanical experiments and managing a plantation of 200 slaves. There in Georgia Joseph and his brother John grew to manhood in the scientific environment of their father and at the same time learned to admire the gentility of the old-school southern temperament. Young Joseph cultivated a strong feeling for the South and its traditions during his youth and enjoyed an association with both Alexander H. Stephens, who tutored him for entrance to Franklin College, and John C. Calhoun.[3]

After finishing college Joseph entered the College of Physicians and Surgeons in New York, but after completing his studies in 1845 he decided upon a life of science rather than the practice of medicine. In 1850 he entered the first class of the Lawrence Scientific School, the haven of natural science, and sought the careful tutelage of Louis Agassiz. As Agassiz's student, he fell in with a group of young naturalists, among whom were Alpheus Hyatt, Edward S. Morse, Frederick W. Putnam, Alpheus S. Packard, Samuel H. Scudder, Nathaniel S. Shaler, and Addison E. Verrill. While in Cambridge he was also drawn to the constellation of Asa Gray, Arnold Guyot, Oliver Wendell Holmes, Henry W. Longfellow, and James Russell Lowell.[4] In 1852 LeConte accepted a chair of natural science at Oglethorpe University in Milledgeville, Georgia, then took a professorship of geology and natural history at the University of Georgia in Athens, and finally a chair of chemistry and geology at South Carolina College in Columbia. When war broke out, LeConte worked as a chemist, first in the manufacture of medicines and later in the manufacture of munitions at the Niter and Mining Bureau at Columbia, where his brother was superintendent.

After the war both he and his brother resumed teaching positions with the University of South Carolina. However, the effects

[3] Marcus Benjamin, "Joseph LeConte," *Scientific American*, LXVII (Aug., 1892), 133–34; William Rader, "Joseph LeConte: The American Evolutionist and Teacher," *Outlook*, LVI (Aug., 1897), 836–39.

[4] LeConte, Lewis Jones, David A. Wells, and John D. Runkle formed the first graduating class of the Lawrence Scientific School. See Armes, ed., *Autobiography of Joseph LeConte*, 142; Andrew C. Lawson, "Joseph LeConte," *Science*, n.s., XIV (Aug., 1901), 273–75.

of war on the educational institutions of the state had been devastating, and, when invited to join the staff of the new University of California, both brothers accepted. Their appointments, wrote Josiah Royce, an early student of Joseph LeConte, were "in part due to the influence of the political reaction, which swept over California in the years following the war, and which gave the State for a time to the Democratic party, despite its record as a decisively Republican State during the war."[5]

LeConte's work in natural science ranged from the phenomenon of binocular vision to studies of geology, the education of women, the functions of the liver and the larynx, and the problems of flight. His geological activities included speculation on the action of glaciers, lava flow, mountain structure, ore deposits, seismology, and coral growth. His *Elements of Geology*, published in 1878, became one of the most widely used textbooks for schools and colleges, and his prominence in science brought him recognition and membership in the leading scientific societies of the country. From the American Institute of Mining Engineers to the California Academy of Sciences, from the National Academy of Sciences to the Brooklyn Ethical Association, his contributions to the intellectual life of the country reflected his deep philosophical belief that evolution was a principle that ran through the entire realm of nature. Nature was an unbroken chain "from the inorganic and dead through the organic and living up to the intellectual and moral."[6]

As an evolutionary idealist and popularizer of science for the community, LeConte gave to science an artistic charm. He believed that the "logic of science," temporarily concealed by uncertainties and inadequate data, would eventually emerge as a completed masterpiece, enabling the scientist not only to trace the course of man's wisdom from the past but also to analyze and direct his potential in the future. LeConte had an artist's concern

[5] Josiah Royce, "Joseph LeConte," *International Quarterly*, IV (1901), 326–27; E. M. Coulter, "Why John and Joseph LeConte Left the University of Georgia, 1855–1856," *Georgia Historical Quarterly*, LIII (Mar., 1969), 17–40; LeConte to Mrs. T. Sumner, Feb. 21, 1869, *LeConte Family Papers*, American Philosophical Society, MSS.

[6] Joseph LeConte, "Evolution and Human Progress," *Open Court*, V (Apr., 1891), 2780; LeConte, *Evolution and Its Relation to Religious Thought* (New York, 1888).

for unity and therefore never accepted a polarity between man's spiritual and physical nature. Likewise, he never separated science from art—the desire for metaphysical unity, a harmony of all categories of life, was the pervasive theme running through all of his work. Wrote Royce of LeConte, "Every definite series or province of facts or of processes [became for him] a subordinate part of a larger whole, intolerable in its fragmentariness."[7] Art, the highest form of unity, preceded science and was its condition for being. Art led to science, but when science advanced sufficiently it in turn perfected art. Science, the "heavenly daughter of an earthly mother," led men to a fuller understanding of life by uniting empirical data with speculative genius, a union of intuition and patient research.

> Empirical art precedes science and is its condition; rational art comes after science and is its embodiment. Empirical art is the outcome of the use of the intuitive reason, which works without understanding itself, and which in its highest forms we call genius. Scientific art is the outcome of the use of the formal reason which analyzes and understands the principles on which it works. Empirical art may indeed attain great perfection, but sooner or later it reaches its limit and either petrifies or decays. Scientific art, because it understands itself, is of necessity indefinitely progressive.[8]

The highest form of art, the art of government or politics, had existed for centuries and had, of course, preceded the science of sociology. Now, however, the rationalism of sociology counseled in the realm of politics. "Social evolution and the art of government," LeConte wrote, "have now reached a point beyond which they can not go by the use of empirical methods alone."[9] Politics had too often stumbled in the advance of civilization, groping in the dark and even retrogressing after historical crises. The United States, in particular, had begun a disgraceful era of politics as a result of the Civil War. Race problems in the South,

[7] Royce, "Joseph LeConte," 333; "Joseph LeConte," *Dial*, XIII (July, 1892), 81–82; Joseph LeConte, "What Is Life?" *Science*, n.s., XIII (June, 1901), 991–92.

[8] Joseph LeConte, "The Race Problem in the South," Brooklyn Ethical Association, no. 28 (1893), 352; Lawson, "Joseph LeConte," 276.

[9] LeConte, "Race Problem in the South," 353.

destruction of local industries, increase of northern financial power, and fundamental constitutional changes had precipitated a breakdown of representative government and a polarization of interests, sentiments, and passions. It was time for rational methods of government to replace the empirical approach to politics, for only in that manner could man prevent revolutions like the Civil War and its devastating aftermath.[10] National progress could only occur when "science or self-conscious reason" guided the social development of the country. If not, the art of politics in America would decline from historical significance and petrify like the Chinese and Japanese governments.[11]

The loss of property in slaves as a result of the war and the complete collapse of the labor system had brought the South to the brink of prostration. But, more important, it left the South with a race of people unaccustomed to freedom and far lower in the scale of evolution than the Caucasian. The Negro's inferior intellectual and moral position in the scheme of evolution found him unable to fulfill his constitutional right to self-government. While LeConte had defended slavery as a suitable institution for the early education and development of the Negro in America, he nonetheless believed that subsequent race evolution had made the institution "less and less natural, and therefore less and less right." LeConte maintained, however, that the war had been unnecessary. The Negro race had evolved as far as possible under the guidance of slavery, and the South, according to LeConte, had become more and more aware in the years before the war that some degree of freedom would be necessary for further race development. "Slavery would certainly have come to an end," he wrote, "not by the external pressure of a foreign sentiment, but by the internal pressure of race growth."[12]

Still, LeConte argued that southern whites would need some degree of control over the Negro in postwar decades. He suggested that, in place of a master-slave relationship, the state or commu-

[10] *Ibid.*, 357; Armes, ed., *Autobiography of Joseph LeConte*, 238.

[11] Joseph LeConte, "Scientific Relation of Sociology to Biology," *Popular Science Monthly*, XIV (Feb., 1879), 429.

[12] LeConte, "Race Problem in the South," 361.

nity inaugurate a race policy allowing a wider range of freedoms for the Negro but still withholding citizenship and voting rights.[13] In the light of anthropological and medical studies carried out on the Negro during and after the Civil War, LeConte maintained that "control" was all the more necessary because of the Negro's inability to care for himself in freedom, and because race struggle and the increased energy of civilization appeared to threaten his survival.[14] That southern policy would be in conflict with the Constitution was "so much the worse for the fundamental law and the constitutional amendments, for it only shows that these are themselves in conflict with the still more fundamental laws of Nature, which are the laws of God." Just as Massachusetts abolitionists sought justification in the higher law for their illegal activities, so southerners, on behalf of self-preservation and the blood purity of their higher race, subordinated the legal grounds of civil rights legislation.[15]

While the South "controlled" the Negro through newer forms of restrictive legislation, LeConte argued that the South had also to concern itself with the race improvement of the Negro. But education, wrote LeConte, should not be confused with natural race evolution. While education could achieve much, to place major emphasis upon it as a means of race improvement was a delusion. A disciple of Spencer, LeConte believed that while race acquisitions such as education were transmitted, general improvement of the individual was "carried over bodily into the next generation by inheritance, but only in a very small part." Furthermore, the Negro's brightness in youth was not evidence either of educational successes or race inheritance but, rather, of qualities which were present in "nearly all lower races (and, indeed, also of animals)."[16] The Negro's brightness, "quickness of memory, keenness of senses, precocity of perceptive faculties" were not to be confused with "the reflective, originating, rational faculties which develop late, and show themselves in active life rather than in

13 *Ibid.*
14 *Ibid.*, 362.
15 *Ibid.*, 364.
16 *Ibid.*, 365–66; LeConte, "Evolution and Human Progress," 2782–83.

school." Essentially, LeConte felt that the Negro race was still in its childhood, not yet capable of self-government and surely not fit for citizenship. The Negro's progress in the last several centuries had been due less to his own race achievement than to his contact with the Caucasian—a condition of progress acquired through imitation more than natural development. Wherever whites predominated in the South, the Negro population, in imitation of them, acquired the tools of industry and citizenship. Wherever their own numbers were in excess of the whites, "the community [was] essentially African" and the blacks rapidly retrogressed "into savagery, and even resum[ed] many of their original pagan rites and superstitions."[17]

LeConte's concern for race mixture caused him to write two particularly significant articles, "The Genesis of Sex" in *Popular Science Monthly* (November, 1879) and "Effect of Mixture of Races on Human Progress" in the *Berkeley Quarterly* (April, 1880). He agreed with many scientists that a mixture of diverse race qualities had benefited offspring since it hastened evolution through the process of differentiation. But LeConte cautioned that there were natural limitations affecting differentiation in individuals. Like Nathaniel Shaler, he divided genus *Homo* into several distinct species and suggested that the union of two "species" was alien to nature, as proved not only by limited sexual attractiveness but also by limited fertility of the offspring:[18] "In some there is offspring, but the offspring is a sterile hybrid which dies without issue. In some the hybrid is fertile, but its offspring is feeble, and therefore quickly eliminated in the struggle for life with the pure stock, and becomes extinct in a few generations; or else it is more fertile with the pure stock than with other hybrids, and therefore is absorbed into one or other of the parent stocks, and the original species remain distinct. If this were not so, there would be no such thing as species at all."[19]

The question of what degree of differentiation among individ-

[17] LeConte, "Race Problem in the South," 366–67; Armes, ed., *Autobiography of Joseph LeConte*, 234–35.

[18] Joseph LeConte, "The Effect of Mixture of Races on Human Progress," *Berkeley Quarterly*, I (Apr., 1880), 84–85, 89–90.

[19] LeConte, "Race Problem in the South," 369–70.

uals produced good results, and at what point the bad effects would begin, weighed heavily in LeConte's writings. It appeared to him that the best results were found in the crossing of "national varieties, and perhaps of all varieties within the limits of the four or five primary races," but that "the crossing of these primary races themselves produces bad effects." The manifestations of mental and physiological differences separating the primary races were natural barriers beyond which crossing had a bad effect.[20] The product of primary crossing was a hybrid, by nature weaker than either of the two parent races. Feebler and more susceptible to disease, hybrids tended to die out or be reabsorbed into one of the parent races. While believing that the mulatto was intellectually superior to the full black, LeConte, like the medical profession in general, doubted that the mulatto retained the physical capacity of either the Caucasian or the Negro race. Building upon the earlier conclusions of Josiah C. Nott, George R. Gliddon, and Sanford B. Hunt, the California scientist argued that the hybrid mulatto could not maintain himself in a laissez-faire environment. The mixing of these two primary races produced an inferior breed which "must eventually perish" in the natural course of race struggle.[21]

Beginning with the premise that mixtures of primary races produced inferior varieties unable to maintain themselves, and that the lower races seemed already doomed by the laws of race struggle, LeConte reduced the problem to a choice of alternatives. The Negro race must either mix its blood with the higher race or face extermination in the struggle of the fittest. For LeConte, both alternatives were morally unacceptable. In the first place, the natural antagonism of the races would prevent miscegenation. Furthermore, the inferior quality of the mulatto, with physiological features substandard to both the full black and the white, would only hasten race deterioration. On the other hand, the thought of the Negro's extermination through natural race struggle seemed equally deplorable in a civilized society. The institution of slavery had prevented this race extermination by creating an artificial

[20] *Ibid.*, 372; LeConte, *Evolution and Its Relation to Religious Thought*, 226.
[21] LeConte, "Effect of Mixture of Races," 100–101.

buffer between the two races. Forced to live in an artificial position outside the field of race struggle, the black race, rather than succumbing to race struggle like the American Indian, absorbed the Caucasian's civilization and maintained a stable position in an otherwise impossible environment. Their "compulsory service . . . in return for protection and guidance" was far more humane than "neglect and consequent extermination." Only by perpetuating this paternalism in the postwar decades could white society protect the Negro from extermination.[22]

Rejecting both alternatives as too extreme, LeConte's evolutionary idealism sought a middle ground, attempting to find the solution somewhere in a synthesis of the two. The race problem in the United States, he argued, was different from that of other countries. The "fair-haired Teuton" and the Negro represented extremes in the hierarchy of the races.[23] However, borrowing an idea from David G. Croly, author of *Miscegenation: The Theory of the Blending of the Races, Applied to the American White Man and Negro* (1864), LeConte suggested that if a solution was to be found on a humanitarian basis for the preservation of the inferior races, it would have to be in the "judicious crossing" of the "marginal varieties" of Caucasian with inferior races.[24]

Reflecting a "liberal" Victorian outlook, LeConte maintained that "judicious crossing" was the only real alternative to race extinction. By removing the Teuton from direct confrontation with the Negro, he placed the obligation for race mixture on the "inferior varieties" of the Caucasian race. Without directly mentioning Croly's suggestion of a "melaleuketic union" between the Irish and the Negro, LeConte argued that in order to achieve the "perfect ideal humanity" of the future, the Teuton should permit a carefully selected and judicious mixture of intermediate stocks without touching or sacrificing the blood of his own superior stock.[25] Since diversity within the primary races had produced

[22] LeConte, "Race Problem in the South," 360.

[23] *Ibid.*, 374.

[24] LeConte, "Effect of Mixture of Races," 101.

[25] LeConte, "Race Problem in the South," 374; David G. Croly and George Wakeman, *Miscegenation: The Theory of the Blending of the Races, Applied to the American White Man and Negro* (New York, 1864), 29–30.

an excellent stock and a progressive civilization, the mixture of those highest civilizations with a judicious mixture of the marginal varieties could "produce a generalized type capable of indefinite progress in all directions." Civilization would then no longer be Teuton, Aryan, or Caucasian but, rather, "human." Thus the civilization of the future would "be coextensive with human nature, with the earth surface, and with the life of humanity."[26] Unless this was done, the lower races were certainly doomed to extinction.[27]

While LeConte believed that judicious mixture was the most humanitarian solution to the race problem in America, he fell back upon Spencer's *Principles of Psychology* to show that races progressed through slow but successively higher physiological and psychological stages. The "ideal" solution required a slow process of evolution and the careful scientific application of rules and legislation reaching toward that solution. Because of this, LeConte thought it important to look with urgency to the more immediate problem at hand in the South. The dilemma of the "solid South" had to be faced and steps taken to break it.[28] With this in mind, LeConte's "progressivism" fell with full force upon the Negro, for he blamed the solid South on the Negro's claim to the ballot and to the channels of government. "The South is not solid against the North or against any party as a party, but she is solid for self-government by the white race as the only self-governing race. Until some better line be drawn defining a self-governing class, she is obliged to be solid. That some such better line be made I can not doubt, for the color-line pure and simple can not continue. It is not only manifestly unjust, and therefore debauching to the political honesty of the whites, but is a constant source of irritation, and therefore fraught with danger."[29] Ironically, LeConte's progressivism suffered the same fate as state progressivism at the turn of the century—the immediate concern for a healthy two-party system depended upon the negation of the Negro's consti-

[26] LeConte, "Race Problem in the South," 375.

[27] LeConte, "Effect of Mixture of Races," 102; LeConte, "Instinct and Intelligence," *Popular Science Monthly*, VII (Oct., 1875), 664.

[28] LeConte, "Race Problem in the South," 375.

[29] *Ibid.*, 376.

tutional rights and the prostitution of the "scientific method" on behalf of race and class interests. In the light of evolutionary idealism and the jargon of scientific sociology, LeConte attempted to rationalize bourbon acquiescence to populist demands in the 1890's that black voters be disfranchised. By eliminating the Negro from the area of politics through educational and property qualifications, LeConte believed that a major step would be taken to maintain self-government for the white community and return the South to a viable two-party system. He emphasized the property qualification over that of education because book education, "easily acquired by the Negro on account of his quick apprehensiveness, had little effect on character, and is but small guarantee for self-governing society."[30]

LeConte's "scientific" explanation for race segregation was no less ominous in its implications. He explained segregation by first relating it to efforts to understand physical nature—the scientist's use of artificial classification to group nature into discernible categories for study and understanding. Because artificial classification was the naturalist's tool for managing the material before him, "any classification [was] better than none; any kind of order [was] better than chaos."[31] With the aid of the comparative method, LeConte carried the analogy into race relations. Inasmuch as social scientists had first to understand human society before they could direct it without chaos, they were compelled, for the sake of clarity, to create classifications of men. Unlike such artificial classifications as wealth, status, and class, race classifications were both natural and rational divisions of society. They were founded on a "real natural difference—a difference in the grade of evolution." They were not only a rational division made to understand the full functions and capacities of the races but were also natural divisions and, therefore, should "not be broken down . . . until we understand better than we now do the laws of the effects of race-mixture. . . . If the effects of the mixture of the extreme primary races be bad, not only immediately, but for all time and under any mode of regulation, then the law of organic

[30] *Ibid.*, 377.
[31] *Ibid.*, 379–80.

evolution, the law of destruction of the lower races and the survival of only the higher, must prevail and the race-line must never be broken over. If, on the other hand, mixture of the extreme primary races can in any way and by any rational mode of regulation be made to elevate the human race, then the race-line must and ought to be broken down and complete mixture must eventually take place."[32]

It was LeConte's belief that the scientist, with his understanding of evolution and his tools of exact methodology, held the keys to proper race relations in America. Physiologists had made startling discoveries in the area of mental phenomena. The "existence of chemical and molecular changes in the brain corresponding to changes of mental states" opened unlimited opportunity for the scientist to measure quantitative differences among individuals and races. By localizing and understanding the faculties of the mind, he wrote, it was possible to lay the foundations of a "truly scientific phrenology."[33] By means of the comparative method social scientists could enlarge the applications of science to the demands of modern problems. Utilizing the truths of the sister sciences and adapting them to the human and social organism, the educational disciplines of the future could advance human possibilities to yet higher levels.[34] The achievement of the ideal man in the ideal society required that "individual interests . . . be subordinated to social interests, but only because society is the greater organism."

But when we remember that human society is an association of individuals not long since emerged out of animality, nor far on the way toward a true, i.e., an ideal humanity, and that the achievement of that ideal is the real end and meaning of our earthly life; and finally, that an organized society is the necessary and only means whereby the ideal may be achieved, whether in the individual or in the race, we see at once that the immediate individual interest must be subordinate to this, the highest interest of humanity. But subordination is not sacrifice. On the

[32] *Ibid.*, 382.

[33] Joseph LeConte, "Man's Place in Nature," *Princeton Review*, 4th ser., II (Nov., 1878), 789.

[34] Joseph LeConte, "The Effect of the Theory of Evolution on Education," National Educational Association, *Proceedings*, XXXIV (1895), 154–55.

contrary, it is the highest success for the individual. In subserving this, the highest interest of humanity, each individual is thereby subserving his own highest interests. In striving to advance the race toward the ideal, he is himself realizing that ideal in his own person.[35]

LeConte, like many theistic Christian evolutionists of the late nineteenth century, expressed a reverence for science that was but a shallow masking of Victorian thinking. His progressivism preserved the customary religious unction, necessarily moral in outlook and meditative over the realization of progress. He had a rational attachment to the scientific method, a speculative admiration for the possibilities of human perfection, yet a primitive suspicion of racial aptitude. Progress became the exclusive monopoly of the Teutonic race, which was steeped in propriety and aware of the partial endowment of intellectual capability in the human animal. LeConte's idealism walked a narrow path, conscious always of limitations of the half-enlightened and determined to leave change in the hands of those best equipped to decipher the variety of worthiness in human nature.

NATHANIEL SOUTHGATE SHALER: THE CRADLE-LAND THEORY

Like many of the supra-organic evolutionists of his age, Shaler browsed with intellectual pleasure, delighted in spirited discussion, and communicated across the broad, if not limitless, boundaries of nature. As one biographer wrote, his life's work exemplified "the naturalist's love of detail and the philosopher's fondness for large problems." He was a compendium of information that encompassed the empirical, the hearsay, and the *a priori*, and as his foundation in science became a stepping-stone into history and philosophy, so it also became a medium through which he explained and "verified" his concepts of race.[36] Though not a tall

[35] *Ibid.*, 158.

[36] "N. S. Shaler," *Science*, XXIII (June, 1906), 871; Walter L. Berg, "Nathaniel S. Shaler: A Critical Study of an Earth Scientist" (Ph.D. dissertation, University of Washington, 1957); Barbara M. Solomon, *Ancestors and Immigrants: A Changing*

man, Shaler gave his students the impression of an ancient Hector or Ulysses. A young admirer once drew a picture of him "standing unmoved in the whirlwind, smiling at the lightnings." His figure assumed a certain stateliness, somewhat like a descendant of a long line of planter aristocrats or perhaps a distinguished preacher from South Church. Mixing anecdote with science, reminiscence with experimentation, he carried forward a systematic study of American science and social science that had implications for the era's most controversial social and political issues. His views about mankind, half scientific, half opinion and recollection, gave him a position of immense authority in the ripening of American attitudes on race.[37]

The Shaler family had migrated from England to Jamaica in the eighteenth century, from there to New York, and then to Connecticut. His father, a graduate from Harvard Medical School in 1828, settled in a slaveholding community in Kentucky where he practiced medicine. His mother, Anne Hind Southgate, came from a prominent Virginia family attached both ideologically and economically to the slave system. Young Shaler followed in his father's footsteps to Harvard, entering the Lawrence Scientific School in 1859 and coming under the tutelage of the famed Louis Agassiz. But it was also at the Lawrence School that Shaler, along with several of his classmates and associates, broke from the cataclysmic theory of Agassiz and sought refuge in the environmentalism of Lamarck. After completing his B.S. degree in 1862, Shaler hurried back to Kentucky where he was commissioned as captain of an artillery company which soon afterward became known as "Shaler's Battery." After resigning his commission in 1864, he returned to Harvard where he worked in the Museum of Comparative Zoology as an assistant to Agassiz in paleontology. In 1866 the young naturalist traveled to Europe where he visited not only the museums and geological landmarks but met with Darwin, Lyell, and Elie de Beaumont. He received a formal appointment

---

New England Tradition (Cambridge, Mass., 1956), 90–94, 104, 173–74, 201–2; Persons, *American Minds*, 289–95.

[37] Frank W. Noxen, "College Professors Who Are Men of Letters," *The Critic*, XCII (Feb., 1903), 127.

as lecturer in paleontology at Harvard in 1869 and a year later was appointed professor. Besides instructing an estimated 7,000 students during his teaching career, Shaler supplemented his university duties with numerous travels and scientific investigations and surveys. During his early years in science he led several field expeditions connected with Harvard, and in 1873 he was appointed state geologist of Kentucky. In 1884 he directed the Atlantic Coast Division of the National Survey, and throughout his long life he published a voluminous amount of material on geology as well as American problems, earning for himself the degree of Ll.D. in 1903 as "naturalist and humanist."[38]

Shaler's interpretation of organic and social evolution weighed heavily in his analysis of the American character or, rather, of the qualities that went into the making of that character. Though he recognized that both Darwin and Weismann had argued against individual "will" determining the shape of those qualities, he remained convinced of the efficacy of human volition in both mental and physical changes and of the transmission of those accomplishments to successive generations. Despite the Darwinian emphasis upon accidental or "incidental" variation through natural selection, there was reason to believe that the theories of Darwin, Weismann, and Lamarck might be drawn together in a higher synthesis.[39]

Nature, like the womb of a mother, enfolded humanity and helped to form it. Man came into the world by numerous stages of development from the lower forms of life, "each stage being attained by the perfect reconciliation of that advancing life with the nature about it." The effects of environment upon the individual brought the human form "forth from the primal chaos and placed its life under the skies of to-day." The creature of the present, though it endured for only a moment in time, was "the heir of all

[38] John E. Wolff, "Memoir of N. S. Shaler," Geographical Society of America, *Bulletin*, XVIII (1906), 598; William H. Hobbs, "N. S. Shaler," Wisconsin Academy of Science, *Transactions*, XV (1904), 924–27; R. W. Dexter, "The Salem Secession of Agassiz Zoologists," Essex Institute, *Historical Collections*, CI (1965), 27–39.

[39] Nathaniel S. Shaler, "American Quality," *International Review*, IV (July, 1901), 49; Shaler, *The Autobiography of Nathaniel Southgate Shaler* (Boston, 1909).

the ages and embodie[d] in its life the experiences of the past."[40] Each land mass had a particular influence on the development of mankind. It was in each cradle-land that the races developed their more permanent qualities. There, during their prehistoric existence, they acquired permanent race characteristics, the "assemblage of physical and mental motives" that reflected their geographic circumstances. There also the races achieved "fixity of race characteristics," peculiarities which they subsequently carried in their later wanderings. Aryan, Moor, African, and Hun were "still to a great extent what their primitive nature made them." Their race quality was part of the rigidity which came to "mature races in the lower life" and determined, to a certain extent, their future race course as well as the "vigor with which they do their appointed work."[41]

In the acclimatization of Europeans to the American continent, what Shaler called the "American-type of man" eventually developed—a person somewhat thinner and more angular than his European cousin, "quicker witted," and "readier to fit himself to circumstances."[42] But while Shaler accepted a certain physiographic influence of the American environment upon the racial traits of Europeans, he felt that the change had been minor when compared to the more permanent race qualities brought by the European to America. How else, he argued, could one explain the failure of the indigenous Indian civilization?[43] Primary race characteristics were the product of an original cradle-land. Subsequent racial qualities, he wrote, "are not always the playthings of climate."[44] Environment was no longer as important in race development as it had been thousands of years ago. As further proof of his hypothesis, he referred to the Negro slaves in America who were "in no wise altered after an exposure for several generations

40 Nathaniel S. Shaler, "African Element in America," *Arena*, II (Nov., 1890), 661; Shaler, ed., *The United States of America*, 2 vols. (New York, 1894), I, 1–2; Shaler, *The Individual: A Study of Life and Death* (New York, 1900), 252–54, 313–14.

41 Nathaniel S. Shaler, "Nature and Man in America," *Scribner's Magazine*, VIII (Sept., 1890), 364.

42 *Ibid.* (Nov., 1890), 649.

43 *Ibid.*, 654.

44 *Ibid.*, 650.

to the very great change in climate, food, and other conditions which they have found." If natural selection in race life was as important now as it had been at one time, the Negro's migration would have been impossible—the race would have either died out or undergone great physical changes in the process of adaptation. Referring to the earlier medical and anthropometric studies of inferior brain convolution and early suture closure in the Negro race, Shaler concluded that physiographic peculiarities conferred on a race during its cradle-land development posed restrictive barriers upon subsequent race development.[45]

According to Shaler, the American continent had been more than adequate in supporting the racial qualities of the northern Europeans who flocked to its shores. "We may reasonably conclude," he wrote, "that it suits the whole Teutonic branch of the Aryan race."[46] While his original state of Kentucky and his adopted home of New England offered the best of the "American type" from both a physical and intellectual point of view, Shaler feared that New England's exposure to foreign immigration portended a doubtful future. He was openly hostile to the changes which industrialism had brought to the ethnic makeup of the region. New England towns with their mixed races of immigrants were a sharp contrast to the stolid New England stock of the countryside, whose lean and "sturdy-looking" farmers were the last of a hearty race of old New Englanders.[47] The only signs of deterioration in the countryside were the few traces of poverty-stricken Irish who had moved onto abandoned farms. "Such stuff," he feared, "will try the digestion of our New England civilization." In the Connecticut Valley large numbers of Canadian French had already taken over factory and railway jobs, and their numbers seemed to be increasing at a rate that would soon wash out the English blood of the old tillers of the soil. But while Shaler lamented the loss from Anglo-Saxon grasp of the Connecticut

[45] Nathaniel S. Shaler, "Environment and Man in New England," *North American Review,* XLXII (June, 1896), 738; Shaler, *The Citizen: A Study of the Individual and the Government* (New York, 1904), 327–28.

[46] Shaler, "Nature and Man in America," 654.

[47] Nathaniel S. Shaler, "The Summer's Journey of a Naturalist," *Atlantic Monthly,* XXXI (June, 1873), 709.

Valley, he rationalized it with the brief remark that it was "better [to] the French than the Irish." "Mingled with the Yankee population," he wrote somewhat encouragingly of the Canadian newcomers, they "became frugal, industrious, even hard-working people, somewhat given to drink and rather immoral, but with none of the shiftlessness which belongs to the Irishman of the same grade."[48]

Shaler believed that the nation had suffered grievously from its policy of allowing immigration of virtually all alien races. The economic greed of American business which had encouraged immigration—and which misguided humanitarianism had justified and perpetuated—had resulted in strikes, talk of socialism, and efforts to unionize workers. Democracy could barely survive in areas where immigrants from southern and eastern Europe lived. Their political ideals, customs, and social heredity were entirely alien to the traditions of the Anglo-Saxon. Shaler feared that with such citizenry the country would develop a race oligarchy composed of permanently inferior and superior race stocks. Widely differentiating race groups, with their inherent and ingrained differences, were incompatible with American democracy.[49]

A similar oligarchy had existed in the South during the days of slavery and its master-servant institutions had formed a democracy suited to the nurture of race despotism. By condemning this despotism, Shaler did not mean that blacks were entitled to a share of government. "The negro has little or no more place in the body politic," he wrote, "than he has in the social system."[50] His ignorance and "general lack of all the instincts of a freeman" not only justified the policy of disfranchisement but made it "imperatively necessary." Shaler hoped by his admonition that the nation would shrink from repeating the same error on a national level. He compared the peasant's threat to the democratic life blood of New England with the Negro's threat to southern institutions, and hoped that from the experiences of the South the nation would

[48] *Ibid.*, 713.

[49] Nathaniel S. Shaler, "European Peasants as Immigrants," *Atlantic Monthly*, LXXI (May, 1893), 649.

[50] *Ibid.*, 647.

sense the threat which immigration presented to democracy. True democracy was incompatible with race oligarchy.[51] Citizenship reflected the ideals of the nation, and these could never be grafted onto the foreigner. Citizenship referred to an almost organic relationship between psychic development and physiographic identity —"the mere forms of the court are idle mummery unless this work has been done" naturally and had evolved systematically.[52]

It was characteristic of both peasant and Negro that in accepting their inferior lot they lacked any larger sense of the nation and of their responsibilities as citizens. Worse still, they were "controlled by habits and traditions" which divided them from the remainder of the nation "as completely as though parted by centuries or by wide seas." Their isolation from the thought and culture of America made it all but "impossible for them to develop any political quality whatsoever." Centuries of almost unconscious existence had done permanent damage to their capacity for profitable environmental development, including participation in democracy. It was rare indeed when "any one born in the peasant caste [showed] much individuality of mind." Like the Negro of the South, the peasant acquiesced in his assigned lot and there he was content to drag along on the coattails of democracy.[53] Like the Negro, too, the peasant had no social or political longings because his particular inheritances and traditions supplied him with none. While the normal citizen saw himself as a potential activist in the social and political sphere and desired to "assail the social leadership" with the idea of furthering the democratic dream of society, both the peasant and the Negro lacked any such awareness. Their "singular uniformity" had so fixed their present and future habits of existence that they were virtually caste laborers distinct from the rest of American society and outside the beneficent influence of its stimulating environment.[54]

What bothered Shaler even more was the Catholicism of the new

[51] *Ibid.*, 649.

[52] *Ibid.*, 647–48; Shaler, *The Citizen*, 1; Shaler, ed., *United States of America*, II, 62.

[53] Shaler, "European Immigrants," 649.

[54] *Ibid.*, 650; Shaler, *The Citizen*, 99–100.

immigrants. Romanism had stamped an indelible mark of inferiority upon the character and institutions of Catholic Europe, and he warned of its impediment to the future of the United States. The Church had siphoned off the best of Europe's talent for some twelve or fifteen centuries. It offered the easiest escape for the peasant's lowly position, and those with ability who entered the religious life were restricted to a life of celibacy. This continual drain of peasant talent led to a steady deterioration of racial stock. Though Shaler admitted to difficulties in the understanding of heredity, there could be little doubt, he felt, that the Church had acted as a destructive influence on the race life of Catholic Europe.[55]

While Shaler concerned himself with the immigrant problem in New England and had, like Fiske, contributed to efforts of the Immigrant Restriction League, he nonetheless focused most of his attention on troubled social and political developments in the postwar South. In many ways Shaler's concern for the Negro typified the change taking place in the New England mind in the late nineteenth century. The decades after the Civil War marked the hiatus of the old New England conscience. New Englanders looked upon the wartime *cause célèbre* as having been ill conceived and blamed abolitionists for expecting too much too quickly. Unlike LeConte, Shaler accepted no justification for slavery, but he felt too much importance had been made of the southerner's sin and not enough of the Negro's place in nature. Curiously, New Englanders, who had previously fought for the Negro and condemned southern racial ideology, approached the South during the 1880's and 1890's as a prodigal son. Both sections of the country, they felt, had a knotty race problem. Nowhere except in New England and in the South did social lines so nearly run with racial lines. It was a curious marriage of New England parochialism with the racial ideology of the sensitive southern mind. New England casuistry sought in the southerner not only an ally to stop the flow of immigrants but an understanding brother with whom it could meditate upon common burdens. In his pursuit of a meeting point

[55] Shaler, "European Immigrants," 651.

Shaler attempted to re-evaluate Reconstruction and the nature of the Negro in postwar America.[56]

Perhaps the greatest wrong wrought upon the blacks in the United States, according to Shaler, was the excessive liberty given to them after the Civil War. During a period when relations between North and South were at the greatest strain, too much had been expected of newly emancipated blacks. The nation erred in not providing blacks "with a minimum of freedom with provision for schooling and a franchise based on education." Political machinations, as well as the popular delusion that the Negro was merely a white man with dark skin and kinks in his hair, led to a complete misunderstanding of the Negro's capacity, and he became a pawn of the "worst political rabble that has ever cursed the land." Carpetbag governments added to the havoc by raising the hopes of the blacks and destroying the long-existing friendly relationship between the two races. Only the overthrow of those governments removed the "danger of war between the blacks and whites in the postwar decade." As the carpetbag governments fell apart in the 1870's, blacks renewed their allegiance to their old masters; yet race relations were strained to the breaking point. The southern white now distrusted the Negro as a potential enemy who had "to be watched lest he will again win a chance to control the state." Disfranchisement was thus designed to remove the potential for such a threat until the Negro showed signs of education and property consciousness. Though Shaler realized that there was an inequality in disfranchisement, still, he thought it better than the terrorism and "tissue ballots" that predominated in Reconstruction states.[57] Southerners were justified in their "temporary"

[56] R. T. Berthoff, "Southern Attitudes toward Immigration, 1865–1914," *Journal of Southern History*, XVII (1951), 328–60; Walter L. Fleming, "Immigration to the Southern States," *Political Science Quarterly*, XX (Sept., 1905), 275–94; Bert J. Loewenberg, "Efforts of the South to Encourage Immigration, 1865–1900," *South Atlantic Quarterly*, XXXIII (1934), 363–85; Barbara M. Solomon, "Intellectual Background of the Immigration Restriction Movement in New England," *New England Quarterly*, XXV (1952), 47–59; Edward Saveth, "Race and Nationalism in American Historiography: The Late Nineteenth Century," *Political Science Quarterly*, LXIV (Sept., 1939), 421–41.

[57] Nathaniel S. Shaler, "The Negro Since the Civil War," *Popular Science Monthly*, LVII (May, 1900), 29–30; Shaler, *The Neighbor: The Natural History of Human Contacts* (Boston, 1904), 189–90.

disfranchisement of blacks, but he pointed out that if they transformed the policy into a permanent political fixture, the logic of the justification would be lost in irrecoverable damage not only to the relationship of the two races but also to the future political relations of North and South. Rather than use the disfranchisement policy as the final answer to the race problem, the South should "develop in the blacks the qualities which may make them safe holders of the franchise, and . . . give that trust to all who become worthy of it." It was insensible, Shaler wrote, to think in terms of permanent disfranchisement or forced deportation to the American tropics. The South needed black labor and its "exodus would mean the commercial ruin of half a dozen great States."[58]

The solution to race relations in America lay in a long-range program to fit the Negro into the role that had been prematurely given him by the Fourteenth and Fifteenth Amendments. The reconstruction governments in the immediate post–Civil War years believed that they could convert the Negro, like the immigrant, into a "truly free man" merely by the use of the ballot. This ideal, however, fell short of the realization that the Negro held a place in evolution far short of the Caucasian evolvement—that some 2,000 years lay between the primitive African and the present position and state of mind of the American citizen. "We mocked the African with the gift of the franchise," Shaler wrote. "We have to begin where we should have begun thirty-five years ago, with measures that are proportionate to the needs—within a system of education that may serve to develop the saving qualities of the race."[59]

Shaler was deeply concerned with the process of Americanizing the African. Unlike the European, the African had not improved as a result of his movement from one hemisphere to the other. The Negro's characteristics had remained fixed; for Shaler, the reason was easily deduced from his naturalist's view of the world: "As a general rule, among animals the higher members of any

[58] Nathaniel S. Shaler, "The Future of the Negro in the South," *Popular Science Monthly,* LVII (June, 1900), 147.
[59] *Ibid.,* 148; Shaler, *The Citizen,* 230.

group seem to be more variable in character than the lower, and offer less resistance to those agents, whatever they may be, which lead towards change or destruction. . . . Again, it may be reasoned that just as the most highly developed breeds are those which are the most difficult to retain in their best shape, so those races of men which are the most civilized are those the most dependent upon the conditions of environment for their maintenance."[60] The physiographic adaptability of northern Europeans was an "acquired capacity" caused by experiences of "several thousand years of continued migrations, as well as the constant change of seasons in the western region where [they] developed." The African, on the other hand, had lived in a uniform climate during the thousands of years of his race existence, a climate which offered little change and stamped him with an "intense race individuality." Though his long existence in the tropics had made him able to withstand the torrid climate of the region, it also contributed to a psychological and physiological resistance to change. Because transplantation of the African race had produced little physical change, intellectual advances would "necessarily be slow, even under the most favorable conditions."[61]

Like many physicians in the late nineteenth century, Shaler became alarmed by the high death and disease rate of emancipated blacks. He suggested that the cause stemmed partly from the lack of a sustaining white influence in the postwar era. Independent of white influence, Negro communities seldom retained those social principles necessary for civilization.[62] "Unless the black population can be quickly lifted to a higher intellectual and moral plane than now characterizes it," those sections of the South where the Negro did have a majority would "be apt to relapse into barbarism." Advance of the Negro in America depended almost entirely "upon his remaining in close contact with the superior race."[63] As for the

[60] Nathaniel S. Shaler, "Mixed Populations of North Carolina," *North American Review,* CXVI (Jan., 1873), 162.

[61] *Ibid.,* 162–63.

[62] Shaler, "African Element in America," 671; Shaler, ed., *United States of America,* II, 612.

[63] Shaler, "Nature and Man in America," 655.

whole black population since emancipation, Shaler suspected that the death rate was not only abnormally high but that the black population was "stationary, or absolutely decreasing in numbers." Emancipation had undone many of the good points of slavery, throwing the race into a struggle for existence in a country where the race struggle was too one-sided. As a result, the Negro was "not likely to multiply . . . save where he secure[d] the protection afforded by a strong social framework which he cannot construct and for the existence [of] which he must depend on the state-building race."[64]

Shaler's concern for the mulatto was reminiscent of the investigations of Sanford Hunt and the questionnaires answered by examining physicians during the Civil War. According to Shaler, most of the mulattoes drifted into southern cities and, though more intelligent than the average black, they were "unfitted and indisposed to hard labor." Because of their intellectual talent, they took over the professions of barbers, hotel waiters, and house servants and drew to those professions a greater amount of vice than ever before. Furthermore, the mulatto took precisely those positions which the pure black required for his self-development. The hybrid mulatto was not only an obstacle to the progress of the black race, but science had shown him to be physiologically inferior to both parent stocks. If cities provided one of the ideal places where pure blacks "by contact with the white race" might improve their position, then the mulatto's position in the cities must be reconsidered. The only real hope in the race future of the black man lay in the fact that miscegenation between the races had declined since the war and, since the half-breed was "short-lived and unfruitful," Shaler hoped that it would soon die out and leave the South to the two pure-blooded races. "The mulatto, like the man of most mixed races, is peculiarly inflammable material. From the white he inherits a refinement unfitting him for all work which has not a certain delicacy about it; from the black, a laxity of morals which, whether it be the result of innate incapacity for

[64] Shaler, "African Element in America," 671; Shaler, *The Citizen*, 248; Shaler, *The Neighbor*, 136.

certain forms of moral culture or the result of an utter want of training in this direction, is still unquestionably a negro characteristic."[65]

As a nineteenth-century humanist Shaler believed that the solution to race problems in the United States required a generous spirit, but he was inclined to feel that a spirit tempered by scientific study of the problem of human relations was more appropriate. Men of science had an obligation to unravel the mysteries of human nature and human relations. The laws of inheritance, as understood by modern biology and sociology, opened a whole new view of the extent to which man could study race relationships. A proper understanding of human capacity was sufficient justification for the scientist's role in helping to bring an answer to American political and social problems. The main steps in the inquiry would be first, the history of the race, second, the present condition "from the point of view of anthropology, including psychology," and third, "the social and civic quality of the race both in itself and in relation to the white people." Evidence gained from the inquiry would be placed in the "hands of those best fitted to attend to it." Society would then pass legislation specifically to grapple with the problem.[66]

In pursuing this inquiry Shaler pointed out that the Negro was "nearer to the anthropoid or pre-human ancestry of men" than any of the other races, a situation which meant that the Negro had "been much longer in about the state in which we now find him." Like earlier polygenists Josiah C. Nott and George R. Gliddon, Shaler drew his immediate proof from Egyptian monuments which characterized the African as a slave with a marked physiognomy. Evidence from the past served "to attest the existence of the negro in substantially the same shape in which we now find him." In looking at the Negro's social relationships Shaler argued that few had removed themselves from the lowest forms of savagery and in fact showed "no clear signs of ability to climb

[65] Nathaniel S. Shaler, "An Ex-Southerner in South Carolina," *Atlantic Monthly,* XXVI (July, 1870), 57; Shaler, *The Neighbor,* 60–61.

[66] Nathaniel S. Shaler, "Science and the African Problem," *Atlantic Monthly,* LXVI (July, 1890), 38.

the next round of the ladder." They had acquired the habit of subjection to superior peoples throughout their race history, and they borrowed or imitated whatever they needed without struggling through the formal aspects of building an educational, literary, and religious foundation. What institutions they had were superficial imitations of higher cultures.[67] When Africans were brought to America, their "simple yet strongly inherited motives remained with them, undergoing such changes of adjustment, but not in nature, as the exigencies demanded." They accommodated to the institutions of their white masters. The slave owner, for all practical purposes, took the place of the chief "to whom the black for immemorial ages had been accustomed to render the obedience and loyalty which fear inspires." Under this relationship the black slave continued in nearly the same status as before, but his new master, as a representative of a superior race, had inculcated a higher sense of motive in the subjugated race.[68]

Because of the imitative nature of the Negro, the manners of the dominant race made great inroads on race character. The Negro became a good laborer, content with the paternal relationship offered him by the dominant race. The southern slave owner provided Negroes with schooling "such as no savages had ever received from a superior race," and it was "unlikely that a lowly people [would] ever again secure such effective training."[69] The chief products of this American training were the Negro's marked "gentleness and decency," though it seemed clear to Shaler that, with respect to intellectual advance, there was hardly much to be said or to be measured. Nor had there been any development of a business sense or even a feeling of companionship among Negroes. These handicaps, according to Shaler, were due to the retention of tribal instincts which corresponded to the level of physical and cultural evolvement. Then, too, there was no sense of the sacredness of marriage. For Shaler, the Negro had not yet developed the "family" concept.

67 Nathaniel S. Shaler, "The Nature of the Negro," *Arena*, III (Dec., 1890), 25–26; Shaler, *The Neighbor*, 139.

68 Shaler, "Nature of the Negro," 28.

69 *Ibid.*, 31–32; Shaler, *The Neighbor*, 142–43.

In a long and intimate connection with this folk, I have never heard a man refer to his grandfather, and any reference to their parents is rare. The negro must be provided with these motives of the household; he must be made faithful to the marriage bond, and taught the sense of ancestry. This, it is plain, is a difficult task to accomplish, for the reason that the regard for the forefathers was mainly developed in a state of society through which the negro did not pass, and to which he cannot be subjected. It came from a time when, as in the feudal period, men inherited privileges as they do not in our present commonwealth. Marital faith, however, may be inculcated by social laws, and the ancestral sense may possibly be reinforced and extended by the diffusion of knowledge concerning the laws of heredity. It is difficult to see how we can assist the blacks in this perplexing question but it is clearly one of the points where they most need help.[70]

Shaler suspected at the outset that the predominant proportion of blacks in the United States were from the Guinea coast region and therefore were the least capable of the African peoples.[71] Their facial expression, with protruding jaw, was a blend of the human look and a "remnant of the ancient animal who had not yet come to the careful stage of life." Limited in intellectual abilities, the Guinea type was generally "nimble-witted," with a body which dominated the mind. There was also evidence of a rarer Zulu type in the United States. The men of this stock, however, had "a higher and in every way better head." The Zulu was not the sly "child-animal" characterized by the Guinea type but, rather, represented a "vigorous, brave, alert" stock. The Zulu, wrote Shaler, was fit "for anything that the ordinary men of our own race can do." In Virginia there were also examples of the Semitic Negro, of high quality with "a rather tall, lean form, a slender neck, a high head, and a thin face, usually with a nose of better form than is commonly found, sometimes approaching the aquiline." Their straighter hair and facial profile suggested a mixture of Arabian blood; they were a rare stock in America and usually were retained as household servants during the days of slavery.[72]

[70] Shaler, "Nature of the Negro," 34.

[71] Shaler, "Science and the African Problem," 38.

[72] Nathaniel S. Shaler, "Our Negro Types," *Current Literature*, XXIX (July, 1900), 45.

In seeking scientific answers to the race problem, Shaler hoped that future anthropological investigations would attempt to compare the mental and physical condition of the American Negro with his African ancestors. He suggested that anthropometrical studies be made in various geographic sections of the United States to determine just what the effect of climatic variations had been. The study could also deal with disease and longevity, to determine whether the Negro was "relatively less liable to certain forms of disease than the whites, and . . . more open to invasions of other maladies than the European races." What the anthropometrical studies could best ascertain, however, was the extent of mental development of the blacks. Shaler's suggestion was not original. He merely reinforced an earlier plea by Sanford Hunt to carry out autopsies in the postwar decades on both poor whites and blacks in order to ascertain the effects of environment on brain weight.[73] "It would be interesting to know, as we well might expect to from this investigation, whether the brain of the American African is larger than that of his African prototypes: but it would be still more interesting to know whether his capacity for education is greater than that of his savage kinsmen. . . . [I]t may be possible to determine if the two centuries of enforced labor and civilizing influences to which our American blacks have been exposed have had any effect on their mental development."[74]

Education appeared ostensibly as the key to Shaler's concept of race improvement for the Negro. Yet for him, as for other scientists and social scientists of his day, the educational process necessary for the black man's citizenship was not meant to correspond to the white man's educational status. The Caucasian had shaped his own education over many centuries; it had slowly evolved as the product of his own peculiar race experience. Beginning with the "art of continually laboring," he moved forward step by step into the industrial arts and finally to the "development of the commercial sense with the enlargement of view it gives, and from this [to] the common sense of public affairs that makes a democracy

[73] *Ibid.*, 46–47.
[74] Shaler, "Science and the African Problem," 41.

possible."[75] With this in mind, Shaler borrowed from Spencer's theories on primitive mentality to argue that the black man's education had to correspond to his level of mental evolution. This meant that Negro education would not include the higher scholastic forms. "This is not because I disbelieve in such training," he wrote, "but because it seems futile at the present time to waste efforts in giving these people an education for which they are in general by no means ready—which, if attained, does not afford them a way to a suitable station." Using Spencer's argument that differentiation of perception entailed corresponding physiological and psychological specialization, Shaler maintained that the Negro's efforts to become a physician, lawyer, or clergyman were futile, since the required skills had no "definite relation to [his] capacities." Furthermore, Negroes could not expect to enjoy the range of social opportunities afforded by such professions, since "their employment will have to be essentially with their own people." The Negro had to face the reality of instinctive prejudice from both northerner and southerner, which took the form "of certain rules of intercourse, expressing about the same feeling that separates the commissioned officers and the enlisted men of the army."[76]

Only when society saw how thoroughly "exotic" the Negro in America really was, wrote Shaler, could it begin to understand the peculiar problems that existed in the relationship between the races. The difficulties were immense. Under the "covering of imitated manners or stolidity slumber the passions of a mental organization" that differed widely from the Caucasian's. Though there were good qualities in the Negro, there were also elements which made the most humane reformer "almost despair." The "firmest bases for hope we have," argued Shaler, "lie in [the Negro's] strong imitative faculties," faculties which by Shaler's own definition relegated the race to permanent inferiority.[77]

In order for the Negro to become a citizen in the full sense of

[75] Shaler, "Future of the Negro," 148.

[76] *Ibid.*, 154.

[77] Shaler, "American Quality," 60–61; Shaler, "The Transplantation of a Race," *Popular Science Monthly*, LVI (Mar., 1900), 513–24.

the word, he had first to accept the idea of self-help. As long as he looked for "some great external power to lift him to the social and economic level of the whites," wrote Shaler, there was "no chance that he will come to depend upon himself for advancement." The Negro might find his vocation in craft work, but that level was "as high as [he] may be expected to attain." About a third of the black people were fit for the more skilled trades and higher educational training. While the mechanical trades grew more complicated each year, Shaler felt confident that Zulu and Semitic Negroes would qualify for such work and, though they would require more training than the Caucasian worker, they would be proficient and capable workers.[78]

As another alternative to the race development of the black in America, Shaler gave half-hearted support to the idea that American troops stationed in tropical lands be made up of black troops who would be permitted to take their families along with them in order to become "permanently and contentedly established" in places such as Luzon and "elsewhere in the colonies." Though not officer material, the black man made an excellent soldier or at least a good infantry man. Shaler considered supporting American imperialism in non-Aryan physiographic areas by the utilization of what he deemed an inferior race in America. The American flag would follow the Negro into areas of the world not conducive to Aryan blood. The only hazard Shaler saw in the scheme was that it would remove from America those Negroes most capable of self-government. Shaler feared that, like the Catholic Church's debilitating influence on the peasant, an imperialistic policy utilizing black colonizers might further retard the race progress of blacks remaining in the United States. For the present, therefore, he thought it best that the Negro remain in the South, acquire a business sense, save money, and lift himself out of his economic poverty.[79]

While Shaler insisted that the black man should achieve much of this on his own, he also believed that the great number of

[78] Shaler, "Future of the Negro," 148–49; Shaler, *The Citizen*, 232.
[79] Shaler, "Future of the Negro," 150–51.

blacks in the South would continue to seek the protection of the white population. It was an instinctive thing, "the ancient disposition of the weak man to lean upon the strong which has in all ages and lands determined the relations of folk." It was for this very reason that Shaler felt that the South and only the South understood the Negro, and that the improvement of the race, if it was ever to come, had to be achieved without the power of the North or the intervention of the federal government. Federal law had produced the most evil aspects of humanitarianism and philanthropy. For the future, "the only chance for lifting the black man to the full status of the citizen is by leaving his future essentially in the hands of the masterful folk who alone can help him."[80]

Shaler's defense of "lynch law" in the South showed the extreme to which New England casuistry went in attempting to renounce the politics of Reconstruction and to leave the race problem in the hands of southerners. According to Shaler, the American hungered for freedom; his concept of law, which for his ancestors of the Old World "was something in itself to worship," became in the New World "little more than a device for securing freedom." The American valued the ends of law without its sanctity, and thus, while "he is dutifully law-biding, so long as the machine works to his satisfaction as an effective engine of government, he will put it aside to accomplish what seems to him just." Building upon this premise, Shaler judged the South's lynchings as essentially "American" in nature and in no way akin to the excesses of cruelty found in Old World countries.[81] Lynchings were the reaction of a citizenry to a "brutal offense," quite often "an offense against women." Those who avenged the outrage were not the meaner types of society but the "decent men of American, law-abiding type." The particular circumstances of the lynching, therefore, were important to a proper understanding of the psychology of the action. The American citizen detested brutality, yet, when a Negro "outraged and murdered a woman," the citizen reacted differently emotionally from most men who would cry merely for vengeance. To

[80] *Ibid.*, 152; Shaler, "Economic Future of the New South," *Arena*, II (Aug., 1890), 257–68.

[81] Shaler, "American Quality," 61.

the white American, "the wrong is not only against humanity, but against his conception that the other man is like himself." This was the feeling of the southerner participating in a lynching. Nations whose peoples had a sense of law that was "extrinsic" and "to be reverenced as a majesty" had no similar hypothesis upon which they could act, and though they might be shocked by barbarous crime, they "would not be moved swiftly to avenge it."[82]

"I am not in the least disposed to apologize for lynching," wrote Shaler, despite the fact that the deed was "disgusting on the nearer seeing." He felt that one could judge it "a legal execution with the conditions that led to it," and in that way one would see lynching as "not a sign of real lawlessness, nor of a people given to savage outburst of fury." In reality, use of the lynch law was "the mark of a folk in a curious adjustment to their concept of law and of the nature of their fellow men." The men who had been involved in lynchings were "gentle folk" and it was useless to discuss the "dignity of the law" before them since they felt, in actuality, that they were "its most effective agents." Lynching was not a "sign of low moral estate, but rather of a rude though high conception of the measure of protection owed to the defenceless, and above all to women."[83]

Innate race prejudices existed at the basis of all human relations. They were "the brutal gate-keepers of the castle, always ready to sally forth" at the least provocation. Shaler did not object to race prejudices, since without them the higher life of man could not have been achieved. It was because of their strength that the progress of man became a reality. Since race prejudice was real and innate, the relationship between the races was much more complex than most people believed. Foremost in any race analysis had to be a study of animal relations as the first step to human motives and relationships, for it was there that man could observe "how the foundations of the human mind were laid."

Although perhaps not the hundredth part of man's life has been spent in the conditions of man, we seem to see that the greater part of the

[82] *Ibid.*, 62; Shaler, *The Neighbor*, 150–51; Shaler, *The Citizen*, 125.
[83] Shaler, "American Quality," 62.

> human intelligence is that which belongs to the man, and not to his lower kindred. This is doubtless true of that element of the mental process in which distinct ideas are involved; but the fundamental motives, the blind impulses which drive men to actions in which reason takes little or no part, these are not, properly speaking, human qualities at all. They took their shape and attained their power before the human stage of our life began; they have been to a certain extent modified in their action by the development of the higher qualities of the mind, by the growth of the intellect and the expansion of the sympathies, but from their very antiquity they are far more firm-set and self-determined in their action than the higher acquisitions of the mind.[84]

In the realm of man and through the lower races, there was "an almost precise repetition of the conditions of the lower animals." Gathered into tribes, they were generally at war with one another. As the cultures ascended to the higher organizations of human society, they developed the element of sympathy which encompassed peoples outside their own immediate "social station." Charitable motives along with more technical considerations were expressed as the types of man ascended the race scale; yet existing in all were the inherited instincts of the past which, under certain stimulants, emerged in race contacts.[85]

The "frail covering of civilization," warned Shaler, "disappear[ed] in an instant before the strong ancestral passion of rage." This was most commonly seen in the evidences of race prejudices which turned against outsiders. It was an instinctual process that framed the Yorkshire maxim: "There goes a stranger, Bill; 'eave 'alf a brick at 'im." The instinctive hatred that culminated in the Yorkshire remark had been aroused simply "by novelty in the human kind."[86] It stemmed from the failure to recognize oneself in the stranger. The lack of identifying similar motives, feelings, and loves destroyed the element of brotherhood in contact with strange races. Those intellectuals who had gone beyond instinctive traits and had actually befriended the Negro attained a level of culture where sympathy became a duty and where man

84 Nathaniel S. Shaler, "Race Prejudice," *Atlantic Monthly*, LVIII (Oct., 1886), 510.

85 *Ibid.*, 511–12.

86 *Ibid.*, 512–13.

was considered abstractly and apart from innate prejudice. They achieved this understanding through a constructive imagination which overcame instinctive emotions. According to Shaler, this altruism occurred despite the fact that these enlightened individuals, "when closely questioned, confessed to me that they abhorred the sight of [the Negro]; that his black face and other peculiarities of countenance made the most painful impression on their minds."[87]

Shaler's attitudes toward the Negro and the immigrant seldom varied during his lifetime. The instruments of science not only became the foundation upon which he built an imposing bulwark against what he believed to be peoples unassimilative in the composition of the American type, but also concealed attitudes of southern racial paternalism which had matured during his earlier youth in the slaveholding society of Kentucky. While the Anglo-Saxon reaped the benefits of a man-centered evolutionary process, the so-called "inferior races" and "stocks" remained outcasts from the evolutionary struggle, restricted from participation because of innate racial characteristics that were unresponsive to environmental influences.

EDWARD DRINKER COPE AND THE "AMERICAN SCHOOL" OF BIOLOGY

Although the evidence in favor of evolution was overwhelming by the 1880's and 1890's, a growing number of zoologists, entomologists, and paleontologists began to seek alternative explanations for the actual process of evolution. While Darwin and his followers explained evolution in terms of "natural selection" and Spencer by the "instability of the homogeneous" and "survival of the fittest," a group of American scientists, many of them former students of Agassiz, advanced the law of "acceleration and retardation."[88] This latter school of interpretation, which published many

87 *Ibid.*, 514; Shaler, *The Individual*, 164–65; Shaler, ed., *United States of America*, II, 639.

88 Edward Cope, *The Origin of the Fittest: Essays on Evolution* (New York, 1887), 2.

of its dissenting views in the *American Naturalist,* formed around the research and writings of Edward Drinker Cope (1840–1897), Alpheus Spring Packard (1839–1905), and Alpheus Hyatt (1838–1902). Designated by some as the "American School," it was referred to by others as the "Hyatt School," while Packard used the term "Neo-Lamarckism" to clarify its environmentalist position.[89] A large portion of the material upon which their conclusions rested grew from a study of the human species and, in particular, a comparison of the various races. In addition, their conclusions became an added source for middle-class efforts to arrive at a social-scientific solution to the race question.

In 1866 Alpheus Hyatt, later professor of zoology and paleontology and curator of the Boston Society of Natural History, published his first study on genetic relations of fossil cephalopods, and from there moved to investigations of ammonites, mollusks, and sponges. "Convinced that the changes in form and organization in bodily structure in the ammonites were directly correlated with the pressures of their physical environment," Hyatt chose to see environment as the dominant force in species modification. His colleague Alpheus Packard, professor of zoology and geology at Brown University and a descendant of seventeenth-century New England, published an article on cave fauna in the *Bulletin* of the United States Geological Survey (1877) attacking Darwin's principle of natural selection as not adequately explaining the production of new fauna species.[90] Similarly, Edward Cope's initial work on batrachians in 1866 became the introduction to a multitude of studies, including *Origin of the Fittest* (1887) and his exposition of neo-Lamarckism in *The Primary Factors of Organic Evolution* (1896). With men like Joseph Leidy, Edward Cope, Othniel Marsh, and Henry F. Osborn studying vertebrates and James Hall, Alpheus Hyatt, and Charles D. Wallcott investigating the inverte-

[89] Packard first used the term in his *Standard Natural History* (Boston, 1885), iii. See also Lester Ward, "Neo-Darwinism and Neo-Lamarckism," Washington Biological Society, *Proceedings,* VI (Jan., 1891), 53; L. H. Bailey, "Neo-Lamarckism and Neo-Darwinism," *American Naturalist,* XXVIII (Aug., 1894), 661–78.

[90] Theodore D. A. Cockerell, *Biographical Memoir of Alpheus Spring Packard* (Washington, D.C., 1920), 181–82; Francis C. Haber, "Sidelights on American Science Revealed in the Hyatt Autograph Collection," *Maryland Historical Magazine,* XLVI (Dec., 1951), 253.

brate subkingdom, American paleontology took a decided neo-Lamarckian approach to evolution.[91]

In many ways zoologist and paleontologist Cope stood out as the most vocal supporter of neo-Lamarckism. A member of the Pennsylvania Society of Friends, he descended from an old Wiltshire family which had bought land from William Penn in 1687 and prospered from a packet line running between Philadelphia and Liverpool. Edward, the great-grandson of the early shipping magnate, developed a love for nature and science early in his youth, becoming an avid student of Dr. Joseph Leidy at the University of Pennsylvania and spending valuable months in the herpetological collections at the Smithsonian Institution under the close guidance of Spencer F. Baird. During the Civil War Cope's family sent him to Europe where he continued his studies; on returning, he accepted a chair of comparative zoology and botany at Haverford College. In 1878 he bought a part interest in the *American Naturalist* and by 1887 was editor-in-chief. Along with Hyatt and Packard, Cope's American school used the magazine as their scientific forum and slowly whittled away at the Darwinian schema through a multitude of tracts on subjects ranging from mollusks to man.[92] A "theist in philosophy and a creative evolutionist in scientific theory," Cope spent a lifetime in scientific writing, in exploring with the United States Geological Survey, and from 1889 to his death in 1897 as professor of geology and mineralogy at the University of Pennsylvania. Despite his Quaker background Cope had a pugnacious disposition which led sometimes to violent quarrels. Once, for example, in the corridors of the American Philosophical Society, an academic controversy with Persifor Frazer culminated in a frenzied fist fight.[93]

[91] Edward B. Poulton, "Fifty Years of Darwinism," in American Association for the Advancement of Science, *Fifty Years of Darwinism: Modern Aspects of Evolution* (New York, 1909), 9–10.

[92] Marcus Benjamin, "Edward Drinker Cope," in David Starr Jordan, ed., *Leading American Men of Science* (New York, 1910), 313–40.

[93] Henry F. Osborn, *Biographical Memoir of Edward Drinker Cope* (Washington, D.C., 1930), 169; Osborn, *Impressions of Great Naturalists* (New York, 1928), 179; Osborn, *Cope: Master Naturalist* (Princeton, N.J., 1931), 534–35; Persifor Frazer, "The Life and Letters of Edward Drinker Cope," *American Geologist*, XXVI (Aug., 1900), 93–94.

Cope's interests were ubiquitous—from women's suffrage and marriage to extinct reptiles, from the castatomoid fauna of North Carolina to fishes, from general morphology of vertebrates to the Negro and disfranchisement. A master of comparative anatomy, he exerted an immense influence on subsequent investigations in taxonomy and paleontology.[94] With his principal work, centering around fossil amphibians, Cope constructed the nucleus of the neo-Lamarckian school among American evolutionists. He suggested that there were two modes of species development. The first consisted of the law of natural selection and the second of Cope's own law of acceleration and retardation. Species, he wrote, developed from pre-existent species "by an inherent tendency to variation, and have been preserved in given directions and repressed in others by the operation of the law of natural selection."[95] In other words, Darwin's law of natural selection was only the first step in a much larger process of development. Natural selection operated by the preservation of the fittest, while retardation and acceleration acted without reference to "fitness" at all. "Instead of being controlled by fitness," the principle of acceleration and retardation was "the controller of fitness."[96] Once change began, it would "follow the laws of acceleration and retardation and could move at a pace more rapid than Darwin allowed." Neo-Lamarckians relegated Darwin's natural selection to a secondary position "in the conviction that an animal's relation to its environment was the primary cause of evolution."[97] According to them, Darwin's elaborate evidence accounted only for the reality of natural selection but made no attempt to explain the origin of

[94] American Philosophical Society, *Addresses in Memory of Edward Drinker Cope* (Philadelphia, 1897), 1–10; Edward Cope, "The Relation of the Sexes to Government," *Popular Science Monthly*, XXXIII (Oct., 1888), 721–30; Cope, "The Marriage Problem," *Open Court*, II (Nov., 1888), 1307–10, 1320–24; Cope, "The Lynching at Paris, Texas," *ibid.*, VII (Mar., 1893), 3606; Frazer, "Life and Letters," 111–12. Osborn's biography of Cope makes no mention of his articles in *Open Court* concerning Negro disfranchisement.

[95] Edward Cope, "On the Origin of Genera," Philadelphia Academy of Natural Sciences, *Proceedings* (1868), 299–300.

[96] *Ibid.*, 244.

[97] Edward J. Pfeifer, "The Reception of Darwinism in the United States, 1859–1888" (Ph.D. dissertation, University of Michigan, 1958), 165; Jules Marcou, *Life, Letters, and Works of Louis Agassiz*, 2 vols. (New York, 1895), II, 100.

variations.[98] Because natural selection was a preservative rather than an originative principle, Packard argued that Darwin's natural selection and Spencer's struggle of the fittest were "misused to state the cause, when they simply express the result of the action of a chain of causes."[99] Recognizing the dual function of natural selection and the "increments of change impressed upon individuals during their lifetime" and perpetuated "in some measure through heredity," neo-Lamarckians attempted a reconciliation of both Darwin and Lamarck.[100]

"We all admit," wrote Cope, "the existence of higher and lower races, the latter being those which we now find to present greater or less approximation to the apes." Referring to the Civil War anthropological investigations by the U.S. Sanitary Commission, Cope indicated that there was conclusive evidence of the Negro's close structural approximation to the anthropoid—the flattening of the nose, the prognathism, the facial angle, "the deficiency of the calf of the leg, and the obliquity of the pelvis." These physiological characteristics were also observable to a degree in certain "immature stages of the Indo-European race," notably among the Irish and Slavic peoples.[101] Here was conclusive proof, he felt, of parallelism and the law of acceleration and retardation. By parallelism he meant that "while all animals [or men] in their embryonic and later growth pass through a number of stages and conditions, some traverse more and others traverse fewer stages." Though the stages were nearly the same in different races, "those which accomplish less resemble or are parallel with the young of those which accomplish more." Although structural characteristics between the two races were similar, the action of acceleration and retardation had created characteristics which were often rudimentary or considerably developed on comparison. The parallelism existing between Caucasian and savage, for example, indicated

[98] Edward Cope, *The Primary Factors of Organic Evolution* (Chicago, 1896), 4–5.

[99] Alpheus Packard, *Lamarck, the Founder of Evolution* (New York, 1901), 391, 402.

[100] Ward, "Neo-Darwinism and Neo-Lamarckism," 12.

[101] Edward Cope, "On the Hypothesis of Evolution," *Lippincott's Magazine*, VI (Aug., 1870), 40–41.

that with the mental advancement of the former there occurred a corresponding retardation in such quadrumanous features as prognathism, facial angle, and dental development. As soon as the Caucasian's mental achievement permitted a greater control over environment and an ability to mold things to his own liking, so his quadrumanous structures underwent significant modification in successive generations. Further intellectual predominance explained his "retarded" molar development in contrast to the jaws of Old World apes and the well-developed dental structures of half-civilized races.[102]

In order to demonstrate more fully the correlation between parallelism and the law of acceleration and retardation, Cope set up an elaborate graph of twenty-three structural evidences of evolution.[103]

I. THE GENERAL FORM

1. the size of the head
2. the squareness or slope of the shoulders
3. the length of the arms
4. the constriction of the waist
5. the width of the hips
6. the length of the leg, principally of the thigh
7. the sizes of the hands and feet
8. the relative sizes of the muscles

II. THE SURFACE

9. the structure of the hair (whether curled or not)
10. the length and position of the hair
11. the size and shape of the nails
12. the smoothness of the skin
13. the color of the skin, hair, and irises

III. THE HEAD AND FACE

14. the relative size of the cerebral to the facial regions
15. the prominence of the forehead
16. the prominence of the superciliary (eyebrow) ridges
17. the prominence of the alveolar borders (jaws)

[102] Edward Cope, "Evolution and Its Consequences," *Penn Monthly Magazine*, III (May, 1872), 230.

[103] Edward Cope, "The Developmental Significance of Human Physiognomy," *American Naturalist*, XVII (June, 1883), 618–19.

18. the prominence and width of the chin
19. the relation of length to width of skull
20. the prominence of the malar (cheek) bones
21. the form of the nose
22. the relative size of the orbits and eyes
23. the size of the mouth and lips

Utilizing these characteristics, Cope first sought to compare the general physiological differences of man to his nearest paleontological relatives, the quadrumana. As to the general form, Cope found that in the apes the arms were relatively longer and the extensor muscles of the legs smaller. He also noted that the anthropoid body was covered with hair which was neither crisp nor woolly, the hair of the head was shorter, and the color of the skin was dark. Turning to the head and face, he concluded that the facial region of the skull was larger than the cerebral region and that the forehead was not as prominent but generally retreating. Further, the superciliary ridges, the edges of the jaws, and the cheekbones were more prominent while the chin was less so. Last, the nose was short with flat cartilages and no bridge, the orbits and eyes were smaller (except in *Nyctipithecus*), and the mouth was small and thin-lipped.[104]

Cope reasoned that the possession of any of the above quadrumanous characteristics in specific individuals or races gave visual evidence not only of evolution but also of racial inferiority. But he did not stop there. He next considered man from an embryological point of view. Taking the original twenty-three structural evidences of evolution, he set up a list of physiological characteristics that separated the adult man from the infant. In the general form the head of the infant was relatively much larger than the adult's, the arms were relatively longer, the legs and especially the thighs were much shorter, and there was no waist. Also the body was covered with fine hair, while that of the head was short. With respect to the head and face, the cerebral part of the skull greatly predominated over the facial, the superciliary ridges were not developed, and alveolar borders and malar bones were not

[104] *Ibid.*

prominent. The nose had no bridge and the cartilages were flat and generally short. Finally, the eyes were larger.[105]

Cope suggested that individuals or races which presented any of the above characteristics were "more infantile or embryonic in those respects" than others but that "those who lack them have left them behind in reaching maturity."[106] On the basis of evidence derived from the graphs, Cope concluded that the embryonic structure of the infant monkey had a strong similarity to the embryological characteristics of man. He also saw a far greater difference between the embryonic monkey and the adult monkey than between the human child and the adult man. While man was, in a sense, more embryonic in his facial development, the monkey, on the other hand, showed a much fuller course of growth. Man was distinguished by a large head, prominent forehead, and short jaws. He also had short canine teeth, short arms, and "thumb of hind foot not opposable." In the monkey the reverse was true. Thus, acceleration and retardation, although not evident in the young, showed up in marked contrast in parallel growth at later stages.[107] By placing the Negro, Mongol, and Indo-European alongside the structural evidences of evolution in both the paleontological and embryological graphs, Cope arrived at the following conclusions.[108]

NEGRO
hair crisp (a special character), short (quadrumanous accelerated)
jaws prognathous (quadrumanous accelerated)
nose flat, without bridge (quadrumanous retarded)
malar bones prominent (quadrumanous accelerated)
beard short (quadrumanous retarded)
arms longer (quadrumanous accelerated)
extensor muscles of legs small (quadrumanous retarded)

MONGOLIAN
hair straight, long (accelerated)
jaws prognathous (quadrumanous retarded)
nose flat or prominent, with or without bridge

105 *Ibid.*
106 *Ibid.*, 621.
107 *Ibid.*, 622–23; Cope, "Evolution and Its Consequences," 235.
108 Cope, "Human Physiognomy," 623–25.

malar bones prominent (quadrumanous accelerated)
beard none (embryonic)
arms shorter (retarded)
extensor muscles of leg ("calf") smaller (quadrumanous retarded)

INDO EUROPEAN hair long (accelerated)
jaws orthognathous (embryonic retarded)
nose (generally) prominent, with bridge (accelerated)
malar bones reduced (retarded)
extensor muscles of the leg large (accelerated)

Both the Negro and the Mongolian appeared to have a predominance of quadrumanous features which were retarded in the more advanced races. Borrowing many of his ideas from Fiske's theory of infancy, first suggested in 1871 and subsequently worked out in *Outlines of Cosmic Philosophy*, Cope argued that the Indo-European stopped earlier than both Negro and Mongolian in respect to facial development and, in a sense, was far more embryonic. Predominance of the forehead and reduced facial angle in the Indo-European were a retardation. On the other hand, the nose with its elevated bridge showed a superaddition or acceleration not evident in either quadrumanous or embryonic structures. The bridge of the nose was due to the "development of the front of the cerebral part of the skull and etmoid bone" and hence was a result of cerebral development. Accelerations in body structure of the Indo-European marked progression away from the quadrumana and those features exemplary of brute capacities. Both their embryonic and accelerated features were indicative of added mental development associated with "a greater predominance of the cerebral part of the skull, increased size of cerebral hemispheres, and greater intellectual power."[109]

With the growth of Indo-European societies and the corresponding complexity of their social relationships, the psychological and physiological constitution of the Caucasian changed, and moral and intellectual force took precedence over sheer brute force. Their large brains, "those with the richest convolutions, and with the most delicate structure, with the most appropriate histological el-

[109] *Ibid.*

ements" remained unequaled in the annals of human progress.[110] Developing out of "a primitive state of inactivity and absolute ignorance," they exhibited a gradual psychological and physiological progression from those capacities peculiar to their neighboring races and more remote simian ancestors. Though their brains scarcely differed structurally from the inferior races or the ape, there was, nonetheless, an essential difference in power and capacity.

> Like water at the temperature of 50 and 53 degrees, where we perceive no difference in essential character, so between the brains of the lower and higher monkeys no difference of function or of intelligence is perceptible. But what a difference do the two degrees of temperature from 33 to 31 degrees produce in water! In like manner the difference between the brain of the higher ape and that of man is accompanied by a difference in function and power, on which man's earthly destiny depends. In development, as with the water, so with the higher ape; some Rubicon has been crossed, some floodgate has been opened, which marks one of Nature's great transitions, such as have been called "expression-points" of progress.[111]

The achievements of modern science had permitted scientists to detect qualities of mind by studying external marks on the human physiognomy. The mind functioned as part of the body and, sharing in its "perfections and defects," also exhibited "parallel types of development." Like John Fiske and Herbert Spencer, who had argued a direct relationship between the mental mass of the higher races and "all the conspicuous physical peculiarities of men," Cope also noted that "every peculiarity of the body has probably some corresponding significance in the mind, and the cause of the former are the remoter causes of the latter."[112] In the evolution of man there were not only many divergent races but also obvious physiological and psychological distinctions because many races did "not reach the elevation of the summit."[113] The races of man, con-

[110] Paul Topinard, *Anthropology* (London, 1878), 529.
[111] Cope, "Human Physiognomy," 618.
[112] *Ibid.*
[113] Edward Cope, "Review of the Modern Doctrine of Evolution," *American Naturalist,* XIV (Apr., 1880), 264.

spicuous by their physical differences, represented the branches of the life tree, and a gradation ranged "all the way from a rivalry of physical force" among the lower races "to a rivalry for the possession of human esteem and affection" in the superior races.[114]

Evidence of new physiological structures or the construction of tissues and organs beyond those of earlier generations had a direct relationship to cerebral development. Since the brain developed by intelligent rather than by accidental use, and since it caused a corresponding acceleration or retardation in physiological construction, a "grade-structure" could be established to express the relation of mental and physical capacities.[115] Thus acceleration in brain tissue and brain activity caused a corresponding retardation of quadrumanous characteristics in the superior races, an obvious step toward further physiological and psychological evolution. The greater the capacity for perceiving and taking advantage of surrounding circumstances, the greater the influence the race would have over its bodily parts. The success of the fittest reflected an "increase and location of growth-force, directed by the will."[116] Structural acceleration and retardation were the products of "the effect of the control over matter exercised by the mind."[117]

Cope accepted Spencer's theory of cerebral development. The intellectual faculties of man and animals were in part inherited and in part acquired from experience. Progress depended upon acquisitions from experience "since inheritance without addition is mere repetition." "If no acquisitions were made," Cope pointed out, "the cerebral organization inherited by animals would continually repeat the form of their actions as unerringly as the nature of the machine gives the character to the movements propagated through its wheels and cranks."[118] Differences between inferior and superior species were in large part due to mental powers derived by inheritance and acquisition through experience. In descending the

114 *Ibid.*, 268; Cope, *Ethical Evolution* (Chicago, 1889), 1, 5, 6.

115 Edward Cope, "The Methods of Creation of Organic Forms," American Philosophical Society, *Proceedings*, XII (1871), 256.

116 *Ibid.*, 259.

117 Cope, "Modern Doctrine of Evolution," 263.

118 Edward Cope, "The Origin of the Will," *Penn Monthly Magazine*, VIII (June, 1877), 246–47.

scale of humanity "the energy and amount of the rational element grows less and less, while the affectional elements change their proportions." Rational sex and benevolent characteristics likewise diminished. It was entirely probable that, with the exception of power and fear, savages were deficient in all the emotions. This deficiency among the lower races in emotional qualities was very similar to "a condition which resembles one of the stages of childhood of the most perfect humanity." Like Spencer, too, Cope believed that the mental faculties developed more rapidly than any other body organ. The brain and nerves, "the most plastic of all tissues," executed through activity the accumulated inheritance of primitive protoplasm and, through immersion in a continuing, changing existence, threw off its "formed matter" in purposeful acquisition of new brain properties. The greatest harm to the mental faculties came from brain idleness. For this reason "the greatest stimulus to exercise of the brain is human society." The city, therefore, flourishing with continual intellectual stimulus and mental encounter, far outweighed "the passive virtues of country life."[119]

Cope reported many of the conclusions of his scientific investigations in *Open Court*, a magazine devoted to the "religion of science." In a series of articles during the 1890's he spoke out against the Negro, advocating both disfranchisement and forced migration. The inferior character of the Negro mind in the scale of evolution made him unfit for American citizenship. Lacking sufficient standards of rationality and morality, his organic constitution resembled an uncompleted evolutionary development, the result of an acceleration and retardation process far remote from the corresponding evolution of the white race. Unlike the superior races, the Negro no longer existed in an evolutionary schema. His physical development exhibited such a predominance of quadrumanous features as to preclude any further mental growth. Though Cope accepted the Lamarckian emphasis on environmental influence, he regarded as a well-known fact that "species-characters are often very permanent." Evolution was not always possible in every circumstance. "Only certain types have been susceptible of evolution

[119] Edward Cope, "Evolutionary Significance of Human Character," *American Naturalist*, XVII (Sept., 1883), 916–18.

in the ages of past time"; most types are "side-tracked and left behind." Having had as much time in the past as all other races to develop his education, the Negro had neither "improved it, nor been improved by it." Reiterating the earlier conclusions of men like Alfred Russel Wallace, John Fiske, Herbert Spencer, and others, Cope saw the Negro as "susceptible of education in his youth, and bright and intelligent to a considerable degree." But with puberty the Negro mind underwent "more or less an eclipse."[120]

Because the Negro race was "inferior in character to the neolithic and most of the paleolithic races of prehistoric Europe," it became political suicide to permit Negroes to utilize their million or more votes in the American electoral system. It opened the ballot box to corrupt machinations and demagogues who would "appeal to the superstitions of the Negro."

> While the negro vote can, of course, not control our government alone, it may do so precisely as the smaller vote of New York City has elected at least one president, and has otherwise seriously impressed itself on the general government. It may readily on numerous occasions hold the balance of power. It may govern directly at least two states, South Carolina and Mississippi, and so send four senators to Washington, and in case of closely drawn issues control the senate. It will be supreme in very many local districts of the South. All this only requires to be mentioned to be understood. Its evils have innumerable ramifications throughout our body politic.[121]

For that reason, Cope advocated a restriction of suffrage rights for both whites and blacks and suggested either property or educational qualifications, or both, as possible alternatives to the lynch law, which he deplored. Along with an amendment to narrow voting rights, he also sought a more restrictive immigration bill. America's republican institutions, he warned, depended upon the high moral and physical character of its people. If they lost the superior intelligence needed to govern themselves through mis-

120 Edward Cope, "Two Perils of the Indo-European," *Open Court,* III (Jan., 1890), 2052–53.

121 Edward Cope, "The African in America," *Open Court,* IV (July, 1890), 2400.

cegenation or loss of voting power, the government would more than likely turn to militarism. The franchise, therefore, ought to be guarded from both the "half-civilized hordes of Europe" and the inferior African race.[122]

The United States had enough difficulty assimilating the immigrants flowing into eastern ports every year to attempt, in addition, "to carry eight millions of dead material in the very centre of our vital organism." "We breed our own poison in the slums of our great cities in sufficient abundance," he argued. Nor should the highest race compromise its accomplishments and posterity by mixing with the black. "It would be a shameful sacrifice, fraught with evil to the entire species. It is an unpardonable sale of a noble birthright for a mess of potage. We cannot cloud or extinguish the fine nervous susceptibility, and the mental force, which cultivation develops in the constitution of the Indo-European, by the fleshy instincts, and dark mind of the African. . . . The greatest danger which flows from the presence of the negro in this country is the certainty of the contamination of the race."[123] Reacting against the supporters of miscegenation, Cope reminded his readers that race mixture would cause a deterioration in the intellectual, moral, and political fiber of the nation. For that reason he favored the bill of Senator John T. Morgan of Alabama which sought "to draw the lines of political separation as clear and as deep as is the line of racial distinction between them."[124] Morgan urged the United States to re-examine the Negro's qualifications to suffrage, control the privilege of voting, and secure, if possible, a "happy home" in the Philippine Archipelago "to which [Negroes] would flock with rejoicings and grow into power beneath our flag."[125]

The Caucasian, because of his much greater mental development, had become a more "idealistic thinker" than the rest of mankind. But Cope warned that Caucasian idealism was a peril to

[122] Edward Cope, "What Is Republicanism?" *Open Court,* X (Apr., 1896), 4899.
[123] Cope, "Two Perils," 2054.
[124] John T. Morgan, *Negro Suffrage in the South* (Washington, D.C., 1900), 12; Morgan, "The Race Question in the South," *Arena,* II (Sept., 1890), 385–98; Morgan, "Shall Negro Minorities Rule?" *Forum,* VI (Feb., 1889), 586–99.
[125] Morgan, *Negro Suffrage in the South,* 16; Edward Cope, "The Return of the Negroes to Africa," *Open Court,* III (Feb., 1890), 2331.

race stamina, particularly when relationships "which are in themselves logical and apparently ethical" conflicted with "our material relations" with the inferior races.[126] Such was the case of the African as citizen and voter in the United States. Caucasian idealism appeared not to recognize the Negro's natural unfitness to exercise the privilege of citizenship. "The case is a new one, and demands some independence of thought for its treatment. So-called human rights appear to come into conflict with questions of physical fact or law. The pure idealist will sustain the former, in spite of the latter; but the wise man knows that he must bow to the latter, and acts accordingly. It seems hard to the idealist that inequities between men exist, yet they do exist and appear to work injustice. But we cannot help it." The form of government adopted by the American people, giving "the greatest amount of personal liberty," was a danger to those inferior races "not equally capable of sustaining this relationship between order and freedom." And the Negro, more conspicuous in his failings than any of the other inferior races, failed in all forms of government save that of absolute government. The Negro's dilemma was all the more difficult in that he had to compete with the highest race. Cope concluded that self-preservation was a far more urgent factor than American idealism in political theory. The African, in spite of his preference for remaining in America, ought to be transported to Africa or elsewhere. "The negroes can be spared," he argued, despite "the supposition that the South is not adapted for white labor."[127]

Like other scientists and social scientists, the neo-Lamarckians worked through the late nineteenth century looking for evidence for the primary factors of evolution. Even though the later Mendelian system of heredity undermined the foundations of their school, making it into a charming specimen of nineteenth-century environmentalism, it would not be too much to say that almost the whole "generation of American paleontology carried out their research by adapting the views and methods of Neo-Lamarckism."[128] By the same token Edward Cope utilized much of the scientific

126 Cope, "Two Perils," 2052.
127 Cope, "Return of the Negroes," 2110.
128 Haber, "Sidelights on American Science," 253.

terminology of the neo-Lamarckians in his political pursuits during the growing race consciousness of the 1890's. Their social-scientific vocabulary became the basis of his rationalizations on race and the means by which his racist thought acquired a sense of scientific certainty. For Cope, neo-Lamarckism explained the framework of American culture. Ironically, it was a culture whose political and social environment, in its own way, helped to dictate and delineate the pattern into which American scientists and social scientists evaluated and explained their findings.

VII *The Politics of "Natural" Extinction*

IN 1902 the American Economic Association published a study by Joseph Alexander Tillinghast entitled "The Negro in Africa and America." Like the earlier work of Frederick Hoffman, Tillinghast's study attempted to bring some sort of synthesis to the century's accumulated evidence of race differences; in particular, he suggested that those characteristics of the American Negro which were most debasing were faults which he shared with his African ancestors and, therefore, were not attributable to the effects of slavery. While the institution of slavery had schooled the Negro in the fundamentals of western civilization, environmental influences had been unable to cope with the overwhelming force of heredity.[1] Those attributes most stereotyped in the black—indolence, carelessness, brutality, deception, and passion—were not the products of American slavery but were uneradicable el-

[1] Joseph Alexander Tillinghast, "The Negro in Africa and America," American Economic Association, *Publications*, III (May, 1902), 136.

ements that formed "an integral part of the West African's nature long before any slavery ever touched our shore."[2] The force of race heredity "obscurely but irresistibly dominat[ed] Negro life at every point," and the environmental influences of slavery were powerless "to set aside a fundamental law of nature."[3] The nearly nine million Negroes in the American population constituted an ethnic group "so distinct from the dominant race," he wrote, that the United States was "threatened with the inability to assimilate them."[4]

Emphasizing that character was the product of both heredity and environment, Tillinghast admitted that "through choice or control of environment, deliberate human agency may accomplish much toward influencing the ultimate compound." Yet, despite this variable in effecting change, man could not deter the factors involved in heredity. "This mysterious force operates in isolated independence," he wrote, "and we cannot touch it." Negro children brought up in civilization and in the African jungle would surely reflect "very divergent results," but not to the extent that their heredity would fundamentally change. "No ethnic group, with its inborn nature moulded for ages in an undisturbed environment, can be radically transformed within ten or twenty generations."[5]

Tillinghast supported his thesis of racial inferiority by citing the ethnological data compiled by eighteenth- and nineteenth-century travelers. His sources included Daniel Brinton's *Race and Peoples* (1890), Augustus H. Keane's *Ethnology* (1896), David Livingston's *Missionary Travels and Researches in South Africa* (1858), Oscar Peschel's *The Races of Man* (1876), Anthony Benezet's *Some Historical Account of Guinea . . .* (1771), and some thirty others. He also relied heavily upon Benjamin Gould's Sanitary Commission reports, Paul Topinard's *Anthropology* (1878), Nott and Gliddon's *Types of Mankind* (1854), and Hoffman's "Race Traits and Tendencies." The evidence of all these works seemed

[2] *Ibid.*, 148.
[3] *Ibid.*, 149.
[4] *Ibid.*, 1.
[5] *Ibid.*, 2.

to imply that the psychic nature of the black race had "never been enlarged and refined by selection in response to a progressive environment" and thus remained "inferior to that of peoples long subjected to the stress and struggle of rapidly advancing standards."[6]

Since the black race in America was unable to harmonize its hereditary instincts with American social organization, it seemed to revert to its African character when under the condition of strain.[7] According to Tillinghast, the evidence of reversion came as no surprise to the "student of evolutionary phenomena." The "magic of education" could do little to change centuries of savage culture.[8]

> It is the hard fate of the transplanted Negro to compete, not with a people of about his own degree of development, but with a race that leads the world in efficiency. This efficiency was reached only through the struggle and sacrifice prescribed by evolutionary law. There are many who believe that a shorter path to greatness exists, since the science of education has been developed. But so long as the powerful conservatism of heredity persists, scarcely admitting of change save through selection of variations, it is to be doubted whether education has the efficiency claimed for it. Time, struggle and sacrifice have always hitherto been required to create a great race. If these are to be enacted of the Negro, he must traverse a long road, not in safe isolation in a country all his own, but in a land filling fast with able, strenuous, and rapidly progressing competitors. Under such circumstances his position can with difficulty be regarded as other than precarious to the last degree.[9]

The race concepts of the nineteenth century culminated in a segregated society and a disfranchised Negro. Just as the Negro's distinctness necessitated his disfranchisement to prevent a solid black vote, so segregation became a justification for rendering him impotent as a social and economic force in America. The Negro, "with scarcely a conscious nervous system," imposed a perilous social problem for the Caucasian.[10] The white American was as

[6] *Ibid.*, 92.
[7] *Ibid.*, 226.
[8] *Ibid.*, 227.
[9] *Ibid.*, 228–29.
[10] Charles F. Withington, "The Perils of Rapid Civilization," *Popular Science*

much involved in convincing himself of the black man's inferiority as he was in making the Negro accept the evidence of white superiority. The Negro, so went the argument, "thought if his child could only read, write and cipher, he would be in every way the equal of the Caucasian." This misconception prevented the black man from recognizing the difference "between a man with only capacity to fill with infinite labor a postal card and one who could reason out the law of gravity or define the principles of electricity."[11] But for the paternalism of the white, the Negro would have degenerated into barbarism. His position in white society, an artificial standing far superior to his mental or moral capacities, was a token of Anglo-Saxon benevolence and was in no way achieved by his own self-determination.[12]

It was the growing opinion of the white race in the late nineteenth century that the Negro, "the pet anxiety" of misapplied philanthrophy, would be emasculated of the virtues of self-reliance. Race improvement achieved *for* the Negro, but not *by* him, would raise his hopes, at the same time exemplifying a breach between the races that could never be filled, and would bestow upon him an imaginary importance that the poor white never had in his effort to lift himself. By means of segregation—the conscious political and social effort to keep the Caucasian and Negro apart—the Negro would be removed from the abusiveness of white philanthropy and would, through his own consciousness and purposeful uplifting, gradually make some improvement in his own

---

Monthly, XXVI (Dec., 1884), 233; Philip A. Bruce, *The Plantation Negro as a Freeman* (New York, 1889), 260; W. S. McCurley, "The Impossibility of Racial Amalgamation," *Arena*, XXI (Apr., 1899), 446–55; J. M. McGovern, "Disfranchisement as a Remedy," *ibid.*, 438–46.

[11] James B. Craighead, "The Future of the Negro in the South," *Popular Science Monthly*, XXVI (Nov., 1884), 41; [Lex], "Negro Education," *American Magazine*, VI (1887), 634–35.

[12] Craighead, "Future of the Negro," 44; Tillinghast, "The Negro in Africa and America," 176, 226; Otis T. Mason, "The Savage Mind in the Presence of Civilization," Anthropological Society of Washington, *Transactions*, I (1892), 45–47; Bishop Potter, "The Help That Harms," *Popular Science Monthly*, LV (Oct., 1899), 721–32; Charles Smith, "Have American Negroes Too Much Liberty?" *Forum*, XVI (Oct., 1893), 176–83; Nathaniel S. Shaler, *The Neighbor: The Natural History of Human Contacts* (Boston, 1904), 158–59, 176–77; James Bryce, *The Relations of the Advanced and the Backward Races of Mankind* (Oxford, 1902), 39.

race. For some white Americans, there was the hope that the necessity of a segregationist society would disappear in time, "swept away by the uplifting of the negro to a plane whence he can prove his title to as high consideration in all respects as his white brother."[13]

The sanction of science for the inferiority of the Negro was so formidable that even some black intellectuals accepted both the evolutionary framework and the apparent fate that awaited the Negro in an unsuccessful struggle. In *Progress of a Race,* a study published in 1898 by Negroes Henry F. Kletzing and William H. Crogman, with an introduction by Booker T. Washington, the authors accepted fully the evidence of Negro mortality and the advancing march of Anglo-Saxon civilization. New Zealanders, Tasmanians, Pacific Islanders, and the Negroes of South Africa "perished, not because of destructive wars and pestilence, but because they were unable to live in the environment of a nineteenth century civilization. . . . Their destruction was not due to a persecution which came to them from without, but to a lack of stamina within. Their extermination was due to the inexorable working out of a law as natural as the law of gravitation. And be it remembered, that these races perished in spite of the humanitarian and philanthropic efforts that were put forth to save them. They perished because they had not power of resistance within."[14]

Accordingly, Kletzing and Crogman argued that the Negro race in America had either to "keep up the procession, or else . . . it has to get out of the way." The world was moving too fast; the race could not just sit by the wayside. Race struggle was a reality and Negroes in America had to face it. "Those of us who cannot keep up with it are bound to be crushed to pieces by it." It was not the duty of the Caucasian to retard his own race progress in order to accommodate to the Negro's inferior abilities. The Negro, further-

[13] J. M. Keating, "Twenty Years of Negro Education," *Popular Science Monthly,* XXVIII (Nov., 1885), 25; D. Kerfoot Shute, "Racial Anatomical Peculiarities," *American Anthropologist,* o.s., IX (Apr., 1896), 131–32; M. Alfred Fouille, "Scientific Philanthropy," *Popular Science Monthly,* XXII (Feb., 1883), 521–35; Franklin Smith, "The Real Problems of Democracy," *ibid.*, LVI (Nov., 1899), 1–13.

[14] H. F. Kletzing and W. H. Crogman, *Progress of a Race* (Atlanta, Ga., 1898), 282.

more, should be too proud even to ask for such accommodation.[15] Similarly, Jeffrey Brackett's "Notes on the Progress of the Colored People of Maryland Since the War" (1890) quoted a Negro lawyer from Baltimore who warned his race of the struggle ahead. The future of the Negro race was at stake, the lawyer said. Those who failed in the struggle for survival lacked "qualities of mind, soul or body." "Small mental powers and the consequent inferior character," he advised, "can no more exist in a free contact with a superior people, than can man live amid the raging Vesuvius."[16]

But just as the ideology of white and black separation posed a solution to the black man's status and put off for the future, and to the black man's initiative, the gradual closing of the gap, so it also offered a palliative to white society grasping at a solution to its own race prejudice, and put off to the future its amelioration. The segregation of society provided a scapegoat for white consciences by confronting the Negro with an awareness of his inferiority and convincing him that improvement, to be meaningful, must come from within; and the time, if and when he chose to do this, was entirely for the Negro to decide. Science and its derivative disciplines agreed that the Negro might begin, like the early Anglo-Saxon immigrant, at the base of society and, upon learning the elementary industrial skills, move gradually up the ladder. "A man educated out of touch with himself," wrote Negro Hugh M. Brown, "is like a poor little David clothed in the mighty armour of Saul."[17] Only the Negro's own self-reliance, achieved through his own accomplishments, could bring him out of his inferior status. Segregation in this sense was accepted by both races, though it obviously was determined initially by the unwillingness of the white American to rid himself of race prejudice.[18]

Yet there was more to the racial assumptions of late nineteenth-century medicine and anthropology than simply the scientific rationalization of disfranchisement and segregation. In a very real sense

[15] *Ibid.*, 283.

[16] Jeffrey R. Brackett, "Notes on the Progress of the Colored People of Maryland Since the War," Johns Hopkins University, *Studies in Historical and Political Science*, 8th ser. (1890), 442.

[17] Hugh M. Brown, "Remarks," *Journal of Social Science*, XXXIV (1896), 96.

[18] Shaler, *The Neighbor*, 176–77.

there were two levels to its scientific ideology. On the one hand, the century's scientists and social scientists were saying that the Negro race could only develop naturally outside the artificial position created for it by white philanthropy. Only in a laissez-faire environment, struggling with other races in the natural order of things, could the Negro develop an identity of his own. On this level the emphasis was upon the Negro's own self-help, lifting himself as he climbed, developing the race instruments of pride and ability in the anticipation of future race equality.[19] On the other hand, nineteenth-century science approached race relations on an entirely different basis, one which significantly distorted, if not wholly destroyed, the earlier rationalization. On this level physicians, anthropologists, and social scientists felt that the Negro, in a natural struggle with superior stocks, would die out, that his capacity to survive alongside a superior civilization was lacking due to deficient physiological materials and brain power. His greater mortality, they felt, precluded any future for the race. Only the hothouse environment of slavery had preserved the race from the rigors of natural laws.

Scientists and social scientists of the late nineteenth century did not see the separation of the races and the Negro's disfranchisement as instruments used for the creation of a new Negro stock, developing progressively out of natural race struggle; rather, they saw the Negro working in harmony with the laws of nature and slowly succumbing to the rigors of competition. While the ideology of separation seemed at times rational to both Negro and white societies due to the need for self-development, the derivative evidence of permanent inferiority as well as the belief in the Negro's greater mortality rate outside of slavery undermined the initial rationalization. Dr. Charles Bacon of Chicago offered the suggestion that white society "help along the process of extinction."[20] "I don't know whether that is approved generally or not," added Dr. Lewis G. Pedigo of Roanoke, Virginia, but "the only hope for the

19 Mary T. Blauvelt, "The Race Problem," *American Journal of Sociology*, VI (Mar., 1901), 672.

20 Charles S. Bacon, "The Race Problem," *Medicine* (Detroit), IX (May, 1903), 342.

southern end of the United States, is just these forces that are tending to exterminate the negro."[21] Segregation and disfranchisement in this sense were not means of achieving eventual equality or for that matter, even complete separation; rather, they were first steps toward preparing the Negro race for its extinction. They were policies of anticipation for a singular white society in America, not a policy of two races working consciously toward ultimate equality. Accumulated evidence of the Negro's inability to survive in a natural order precluded any real relevancy to the former rationalization other than merely a disguised anticipation for a more fundamental hope or belief.

The subject of race inferiority was beyond critical reach in the late nineteenth century. Having accepted science and its exalted doctrinaires, American society betrayed no sentiment, popular or otherwise, that looked to a remodeling of its social or political habits of race. There was neither concealment nor delicacy among its beliefs. "Society," wrote Henry Adams, "offered the profile of a long, straggling caravan, stretching loosely toward the prairies, its few score of leaders far in advance and its millions of immigrants, negroes, and Indians far in the rear, somewhere in archaic time."[22] Nineteenth-century America was a peculiar amalgam of restless millions, confident of the future and untroubled by the possibility of evolutionary dysteleology. Large sections of history slipped by with presumptuous vanity under the guise of survival of the fittest, before a hesitant few began to question the miscarriage of the evolutionary schema. The accumulated upheaval of railroad stock deals, machine-boss politics, Grantism, and financial panics was required to disturb the calm of this tidal slack-water. Yet the most sententious critics of the nineteenth century's concept of the survival of the fittest in a struggle for existence considered the structure of race inferiority as outside the framework of their discussions. Dissent about the character of evolution had little bearing on the concept of race inferiority and much less upon the derivation of its racist ideas.

[21] Thomas W. Murrell, "Syphilis and the American Negro—a Medico-Sociological Study," Medical Society of Virginia, *Transactions* (1909), 172.

[22] Henry Adams, *The Education of Henry Adams* (New York, 1918), 237.

Bibliographical Essay

ALMOST ANY EFFORT to explain those attitudes of race inferiority that developed within the context of science during the late nineteenth century necessarily covers an enormous amount of material. Among the various source books for bibliographical data are William Z. Ripley, *A Selected Bibliography of the Anthropology and Ethnology of Europe* (Boston, 1899); Felix M. Keesing, *Culture Change: An Analysis and Bibliography of Anthropological Sources to 1952* (Stanford, Calif., 1953); and the *Catalogue of Books in the Library of the Anthropological Society of London up to July 1st, 1867* (London, 1867). Of the three, Keesing's bibliography is the most beneficial as it presents a chronological listing of books and an assessment of anthropological thought in each time period. The latter catalogue, a product of the Anthropological Society of London, is significant for the type of materials that provided the basis for the more substantive ethnological theories of the nineteenth century. In addition, there are numerous bibliographies concerning the Negro. Some of the more useful are Erwin K. Welsch, *The Negro in the United States: A Research Guide* (Bloomington, Ill., 1965); Elizabeth W. Miller, *The Negro in America: A Bibliography* (Cambridge, Mass., 1966); and Earl Spangler, *Bibliography of Negro*

History (Minneapolis, 1963). The most comprehensive bibliography is that of Monroe N. Work, *A Bibliography of the Negro in Africa and America* (New York, 1965). Unfortunately, these bibliographies are of little value with respect to medical and somatometric studies. For this information, there is available the *Index Medicus;* the *Index Catalogue of the Library of the Surgeon-General's Office;* Edward Mussey Hartwell, "A Preliminary Report on Anthropometry in the United States," American Statistical Association, *Publications,* n.s., III (1893), 554–86; and "The Physical and Mental Abilities of the American Negro," *Journal of Negro Education,* III (1934).

Background reading in the area of the late nineteenth century should include Herbert W. Schneider, *A History of American Philosophy* (New York, 1946), especially the chapter entitled "Evolution and Human Progress"; I. Woodbridge Riley, *American Thought from Puritanism to Pragmatism* (New York, 1915); Henry S. Commager, *The American Mind: An Interpretation of American Thought and Character Since the 1880's* (New Haven, Conn., 1950); Vernon Louis Parrington, *The Beginnings of Critical Realism in America: 1860–1920,* vol. III of his *Main Currents in American Thought* (New York, 1930); and Paul F. Boller, Jr., *American Thought in Transition: The Impact of Evolutionary Naturalism, 1865–1900* (Chicago, 1969). Boller's book offers an excellent summary of the period with a good chapter on "The Day of the Saxon." Unfortunately, however, Boller regarded Frederick Hoffman's writings as "silly" and failed to recognize the depth of anthropological background from which he drew his race analysis. Hoffman's mortality statistics on the Negro became the basis upon which both the Metropolitan and the Prudential Life Insurance Companies refused to write life policies on blacks in the 1880's. See my article, "Race, Mortality, and Life Insurance: Negro Vital Statistics in the Late Nineteenth Century," *Journal of the History of Medicine and Allied Sciences,* XXV (1970), 247–61. Other important works for the interested student include Merle E. Curti, *The Growth of American Thought* (New York, 1943); Ralph H. Gabriel, *The Course of American Democratic Thought: An Intellectual History Since 1815* (New York, 1940); Stow Persons, *American Minds: A History of Ideas* (New York, 1958) and *Evolutionary Thought in America* (New Haven, Conn., 1950), which he edited. Other books worth reading are Gertrude Himmelfarb, *Darwin and the Darwinian Revolution* (New York, 1959); Richard Hofstadter, *Social Darwinism in American Thought,* rev. ed. (Boston, 1955); Philip Appleman, ed., *Darwin* (New York, 1970), which contains perhaps the best single collection dealing with the impact of Darwin on western thought; Loren C. Eiseley, *Darwin's Century: Evolution and the Men Who Discovered It* (New York, 1958); and Arthur O. Lovejoy, *The Great Chain of Being: A Study of the History of an Idea* (Cambridge, Mass., 1936).

Of course, Charles Darwin's *On the Origin of Species,* reprint of 1st ed. (London, 1950), reprint of 6th ed. (New York, 1909), and his *Descent of Man,* 2nd ed. (London, 1901), are indispensable to the reader and should be supplemented with Philip G. Fothergill, *Historical Aspects of Organic Evolution* (New York, 1953); Francis C. Haber, *The Age of the World: Moses to Darwin* (Baltimore, 1959); Don C. Allen, *The Legend of Noah: Renaissance Rationalism in Art, Science, and Letters* (Urbana, Ill., 1949); Robert E. Clark, *Darwin: Before and After* (London, 1958); American Association for the Advancement of Science, *Fifty Years of Darwinism: Modern Aspects of Evolution* (New York, 1909); Alfred R. Hall, *The Scientific Revolution, 1500–1800* (New York, 1954); Edward Clodd, *Pioneers of Evolution from Thales to Huxley* (New York, 1897); Raphael Meldola, *Evolution, Darwinian and Spencerian* (Oxford, 1910); Henry Adams, *The Education of Henry Adams* (New York, 1918); and David D. Van Tassel and Michael G. Hall, eds., *Science and Society in the United States* (Homewood, Ill., 1966). Other equally important works include Mark H. Haller, *Eugenics: Hereditarian Attitudes in American Thought* (New Brunswick, N.J., 1963); Marshall Clagett, ed., *Critical Problems in the History of Science* (Madison, Wis., 1959); Henry Holt, *Garrulities of an Octogenarian Editor* (Boston, 1923); August Weismann, *The Germ-Plasm: A Theory of Heredity* (New York, 1893); Francis Galton, *Natural Inheritance* (London, 1889), *Hereditary Genius: An Inquiry into Its Laws and Consequences* (London, 1869), and *Inquiries into Human Faculty and Its Development* (London, 1883); Alpheus S. Packard, *Lamarck, the Founder of Evolution: His Life and Work* (New York, 1901); David Starr Jordan, ed., *Leading American Men of Science* (New York, 1910); and Henry F. Osborn, *Impressions of Great Naturalists: Reminiscences of Darwin, Huxley, Balfour, Cope and Others* (New York, 1928).

In addition, there is Walter E. Houghton, *The Victorian Frame of Mind, 1830–1870* (New Haven, Conn., 1957); John W. Burrow, *Evolution and Society: A Study in Victorian Social Theory* (London, 1966); Eva B. Dykes, *The Negro in English Romantic Thought* (Washington, D.C., 1942); Hoxie N. Fairchild, *The Noble Savage: A Study in Romantic Naturalism* (New York, 1928); James M. Baldwin, *Darwin and the Humanities* (Baltimore, 1909); Lewis S. Feuer, *The Scientific Intellectual: The Psychological and Sociological Origins of Modern Science* (New York, 1963); John D. Davies, *Phrenology, Fad and Science: A 19th Century American Crusade* (New Haven, Conn., 1955); George H. Daniels, *American Science in the Age of Jackson* (New York, 1968); Richard H. Shryock, *Medicine in America: Historical Essays* (Baltimore, 1966); Merle E. Curti, *The Social Ideas of American Educators* (New York, 1935); Helen Lane, "Heredity and Environment in American Social Thought, 1900–1929: The Aftermath of Spencer" (Ph.D. dissertation, Columbia University, 1950); Milton Berman, *John Fiske: The Evolution of a Popularizer* (Cambridge, Mass., 1961); Edward Lurie, *Louis Agassiz: A Life in Science* (Chicago, 1960); Henry F. Osborn, *Cope: Master Naturalist* (Princeton, N.J., 1931); A. Hunter Dupree, *Asa Gray, 1810–1888* (Cambridge, Mass., 1959); John B. Bury, *The Idea of Progress: An Inquiry into Its Growth and Origin* (New York, 1932); and Arthur A. Ekirch, *The Idea of Progress in America, 1815–1860* (New York, 1944).

For works dealing with the history of the idea of race, the reader should be acquainted with Louis L. Snyder, *The Idea of Racialism: Its Meaning and History* (Princeton, N.J., 1962) and *Race: A History of Modern Ethnic Theories* (New York, 1939); and M. F. Ashley-Montagu, *The Idea of Race* (Lincoln, Nebr., 1965), *Man's Most Dangerous Myth: The Fallacy of Race* (Cleveland, 1964), *Race, Science and Humanity* (Princeton, N.J., 1963), and *The Concept of Race* (New York, 1964), which he edited. There is also Edward F. Frazier, *Race and Culture Contacts in the Modern World* (New York, 1957); Melville J. Herskovits, *The Myth of the Negro Past* (Boston, 1958) and *The Anthropometry of the American Negro* (New York, 1930); Alfred L. Kroeber, *Anthropology: Biology and Race* (New York, 1948); Robert E. Kuttner, ed., *Race and Modern Science: A Collection of Essays by Biologists, Anthropologists, Sociologists and Psychologists* (New York, 1967); Earl W. Count, ed., *This Is Race: An Anthology Selected from the International Literature on the Races of Man* (New York, 1950) and "The Evolution of the Race Idea in Modern Western Culture during the Period of the Pre-Darwinian Nineteenth Century," New York Academy of Science, *Transactions*, VIII (1946), 139–65; Eric Voeglin, "The Growth of the Race Idea," *Review of Politics*, II (1940), 283–317; Jacques Barzun, *Race: A Study in Superstition* (New York, 1965); Margaret Mead *et al.*, *Science and the Concept of Race* (New York, 1968); Robert D. Simons, *The Colour of the Skin in Human Relations* (New

York, 1961); Royal Anthropological Institute, *Man, Race, and Darwin: A Symposium on Race and Race Relations* (London, 1960); and Peter I. Rose, *The Subject Is Race: Traditional Ideologies and the Teaching of Race Relations* (New York, 1968).

Some excellent studies pertaining to American attitudes of racial inferiority are Winthrop D. Jordan's *White over Black: American Attitudes toward the Negro, 1550–1812* (Chapel Hill, N.C., 1968); William Stanton, *The Leopard's Spots: Scientific Attitudes toward Race in America, 1815–59* (Chicago, 1960); and George W. Stocking, Jr., *Race, Culture, and Evolution: Essays in the History of Anthropology* (New York, 1968). Jordan's work is thoroughly comprehensive. Of particular relevance are those sections entitled "The Bodies of Men: The Negro's Physical Nature," "The Negro Bound by the Chain of Being," and "Erasing Nature's Stamp of Color." Stanton's work is primarily concerned with the origination controversy which raged during the early days of American science and the implications of both the monogenist and polygenist schools in the slavery controversy. George Stocking's work, consisting of a series of critical essays, is certainly one of the most scholarly studies to date on the implications of race in nineteenth-century science and social science. One should also consult Stocking's "American Social Scientists and Race Theory, 1890–1915" (Ph.D. dissertation, University of Pennsylvania, 1960). Broader but equally important studies include August Meier, *Negro Thought in America, 1880–1915: Racial Ideologies in the Age of Booker T. Washington* (Ann Arbor, Mich., 1963); Idus A. Newby, *Jim Crow's Defense: Anti-Negro Thought in America, 1900–1930* (Baton Rouge, La., 1965); Oscar Handlin, *Race and Nationality in American Life* (Boston, 1957); Forrest G. Wood, *Black Scare: The Racist Response to Emancipation and Reconstruction* (Berkeley, Calif., 1968); G. C. White, "Immigration and Assimilation: A Survey of Social Thought and Public Opinion, 1882–1914" (Ph.D. dissertation, University of Pennsylvania, 1952); D. F. Tingley, "The Rise of Racialistic Thinking in the United States in the Nineteenth Century" (Ph.D. dissertation, University of Illinois, 1953); Thomas F. Gossett, *Race: The History of an Idea in America* (Dallas, 1963); Maurice R. Davie, *Negroes in American Society* (New York, 1949); Rayford W. Logan, *The Negro in American Life and Thought: The Nadir, 1877–1901* (New York, 1954); Howard Brotz, ed., *Negro Social and Political Thought, 1850–1920: Representative Texts* (New York, 1966); Philip D. Curtin, ed., *Africa Remembered: Narratives by West Africans from the Era of the Slave Trade* (Madison, Wis., 1967) and *The Image of*

Africa: British Ideas and Action, 1780–1850 (Madison, Wis., 1964); and Gunnar Myrdal, *An American Dilemma: The Negro Problem and Modern Democracy*, 2 vols. (New York, 1944).

There is also Paul Lewinson, *Race, Class and Party: A History of Negro Suffrage and White Politics in the South* (New York, 1932); Wilbur J. Cash, *The Mind of the South* (New York, 1941); Kenneth M. Stampp, *The Era of Reconstruction: America after the Civil War, 1865–1877* (New York, 1965); C. Vann Woodward, *The Strange Career of Jim Crow* (New York, 1955); Charles E. Wynes, ed., *Forgotten Voices: Dissenting Southerners in an Age of Conformity* (Baton Rouge, La., 1967); John Higham, *Strangers in the Land: Patterns of American Nativism, 1860–1925* (New York, 1955); Barbara M. Solomon, *Ancestors and Immigrants: A Changing New England Tradition* (Cambridge, Mass., 1956); Edward D. Baltzell, *The Protestant Establishment: Aristocracy and Caste in America* (New York, 1964); Frantz Fanon, *Black Skin, White Masks* (New York, 1967); Henry Pratt Fairchild, *Race and Nationality as Factors in American Life* (New York, 1947); Calvin C. Hernton, *Sex and Racism in America* (New York, 1965); Samuel J. Holmes, *The Negro's Struggle for Survival: A Study in Human Ecology* (Berkeley, Calif., 1937); Edgar T. Thompson, ed., *Race Relations and the Race Problem: A Definition and an Analysis* (Durham, N.C., 1939); and Howard W. Odum, *Social and Mental Traits of the Negro: Research into the Conditions of the Negro Race in Southern Towns, a Study in Race Traits, Tendencies and Prospects* (New York, 1910) and *Race and Rumors of Race: Challenge to American Crisis* (Chapel Hill, N.C., 1943).

In addition to the above works, the reader is advised to look into the late nineteenth and early twentieth centuries for related materials. Some of the more significant authors who deal with the subject of race include Joseph P. Widney, *Race Life of the Aryan Peoples*, 2 vols. (New York, 1907); John B. Haycraft, *Darwinism and Race Progress* (London, 1900); Arthur de Gobineau, *The Inequality of the Human Races* (New York, 1915); Jean Finot, *Race Prejudice* (New York, 1907); Sidney George Fisher, *The Laws of Race, as Connected with Slavery* (Philadelphia, 1860); Martin R. Delany, *Principia of Ethnology: The Origin of Races and Color* (Philadelphia, 1879); Joseph Deniker, *The Races of Man: An Outline of Anthropology and Ethnology* (New York, 1900); Robert Knox, *The Races of Men: A Fragment* (London, 1850); Paul Broca, *On the Phenomena of Hybridity in the Genus Homo* (London, 1864); Joseph A. Gobineau, *The Moral and Intellectual Diversity of*

Races, with Particular Reference to Their Respective Influence in the Civil and Political History of Mankind (Philadelphia, 1856); William Z. Ripley, *The Races of Europe: A Sociological Study* (New York, 1899); Theophilus E. Scholes, *Glimpses of the Ages: Or, the "Superior" and "Inferior" Races, So-Called, Discussed in the Light of Science and History*, 2 vols. (London, 1905–1908); John Fiske, *John Fiske's Miscellaneous Writings*, 12 vols. (Boston, 1902); R. H. Johnson, *The Physical Degeneracy of the Modern Negro, with Statistics from the Principal Cities Showing His Mortality from 1700–1897* (Brunswick, Ga., n.d.); John H. Van Evrie, *White Supremacy and Negro Subordination: Or, Negroes a Subordinate Race and (So-Called) Slavery Its Normal Condition*, 2nd ed. (New York, 1870); Alexander Winchell, *Preadamites: Or, a Demonstration of the Existence of Men before Adam* (Chicago, 1880); Robert H. Terrell, *A Glance at the Past and Present of the Negro* (Washington, D.C., 1903); Alfred H. Stone, *Studies in the American Race Problem* (New York, 1908); William W. Elwang, *The Negroes of Columbia, Missouri: A Concrete Study of the Race Problem* (Columbia, Mo., 1904); Henry F. Kletzing and William H. Crogman, *Progress of a Race: Or, the Remarkable Advancement of the Afro-American Negro from the Bondage of Slavery, Ignorance and Poverty, to the Freedom of Citizenship, Intelligence, Affluence, Honor and Trust* (Atlanta, Ga., 1898); Arthur MacDonald, *Colored Children: A Psychophysical Study* (Chicago, 1899); Colin MacNair, *The Race Crisis* (Henderson, N.C., 1904); John Ambrise Price, *The Negro, Past, Present, and Future* (New York, 1907); Frederick Douglass, *The Claims of the Negro Ethnologically Considered* (Rochester, N.Y., 1854); Edward B. Reuter, *The Mulatto in the United States, Including a Study of the Role of Mixed-Blood Races throughout the World* (Boston, 1918); Philip A. Bruce, *The Plantation Negro as a Freeman: Observations on His Character, Condition, and Prospects in Virginia* (New York, 1889); Hermann Burmeister, *The Black Man: The Comparative Anatomy and Psychology of the African Negro* (New York, 1853); Moncure D. Conway, *Autobiography, Memories and Experiences*, 2 vols. (Boston, 1904); David G. Croly and George Wakeman, *Miscegenation: The Theory of the Blending of the Races, Applied to the American White Man and Negro* (New York, 1864); Daniel G. Brinton, *Negroes* (Philadelphia, 1891); Edgar G. Murphy, *Problems of the Present South: A Discussion of Certain of the Educational, Industrial and Political Issues in the Southern States* (New York, 1904) and *The White Man and the Negro at the South* (Montgomery, Ala., 1900); Thomas Nelson Page, *The Negro: The Southerner's*

Problem (New York, 1904); Buckner H. Payne, *The Negro: What Is His Ethnological Status?* (Cincinnati, 1867); William B. Smith, *The Color Line: A Brief in Behalf of the Unborn* (New York, 1905); William T. Alexander, *History of the Colored Race in America,* 2nd ed. rev. (New Orleans, 1887); Albert B. Hart, *The Southern South* (New York, 1910); and Hilary A. Herbert, ed., *Why the Solid South? or, Reconstruction and Its Results* (Baltimore, 1890).

As for the history of anthropology, there are any number of excellent works. One might easily begin with Thomas K. Penniman, *A Hundred Years of Anthropology* (London, 1935). Penniman calls the period from Darwin to the rediscovery of Mendel's law of inheritance "The Constructive Period" of anthropology. W. E. Mühlmann, *Geschichte der Anthropologie* (Bonn, 1948); Margaret T. Hodgen, *Early Anthropology in 16th and 17th Centuries* (Philadelphia, 1964); Alfred C. Haddon, *History of Anthropology* (London, 1910); Sol Tax, ed., *Horizons of Anthropology* (Chicago, 1964); Hoffman R. Hays, *From Ape to Angel: An Informal History of Social Anthropology* (New York, 1958); Pañchānana Mitra, *A History of American Anthropology* (Calcutta, 1933); and Alexander Goldenweiser, *History, Psychology, and Culture* (New York, 1933), are good complements to Penniman's work. Goldenweiser gives special treatment to the impact of Spencer and the unilinear evolutionists. See also Robert H. Lowie, *The History of Ethnological Theory* (New York, 1937) and *Introduction to Cultural Anthropology* (New York, 1934); and Frederica de Laguna, ed., *Selected Papers from the American Anthropologist, 1888–1920* (New York, 1960). In Laguna the selection of articles is made in such a fashion as to clearly designate trends in American anthropological development. There are also Franz Boas, *Anthropology* (New York, 1908), *The Mind of Primitive Man* (New York, 1911), and *Race and Democratic Society* (New York, 1945); Felix M. Keesing, *Cultural Anthropology: The Science of Custom* (New York, 1958); Betty Meggers, ed., *Evolution and Anthropology: A Centennial Appraisal* (Washington, D.C., 1959); Ernst Mayr, ed., *The Species Problem* (Washington, D.C., 1957); Robert F. Spencer, ed., *Method and Perspective in Anthropology: Papers in Honor of Wilson D. Wallis* (Minneapolis, 1954); Alfred L. Kroeber and Thomas T. Waterman, ed., *Source Book in Anthropology* (Berkeley, Calif., 1920).

In addition, there are the excellent studies by Roy Harvey Pearce, *The Savages of America: A Study of the Indian and the Idea of Civilization,* rev. ed. (Baltimore, 1965); and John C. Greene, *The Death of*

Adam: Evolution and Its Impact on Western Thought (Ames, Iowa, 1959). Greene also has numerous articles including "Some Early Speculations on the Origin of Human Races," *American Anthropologist*, n.s., LVI (1954), 31–41, and "The American Debate on the Negro's Place in Nature, 1780–1815," *Journal of the History of Ideas*, XV (1954), 384–96. Good complements to Greene's articles are Edward Lurie, "Louis Agassiz and the Races of Man," *Isis*, XLV (1954), 227–42; and Herbert H. Odum, "Generalizations on Race in 19th Century Anthropology," *Isis*, LVIII (1967), 5–18. Besides the above, the student is urged to read Juan Comas, *Manual de anthropología física* (Mexico City, 1957); Harry L. Shapiro, "The History and Development of Physical Anthropology," *American Anthropologist*, n.s., LXI (1959), 371–79; June Helm MacNeish, ed., *Pioneers of American Anthropology: The Uses of Biography* (Seattle, 1966); Alfred L. Kroeber, ed., *Anthropology Today: An Encyclopedic Inventory* (Chicago, 1953); and Melville J. Herskovits, *Man and His Works: The Science of Cultural Anthropology* (New York, 1948). The "Symposium on the History of Anthropology," *American Anthropologist*, n.s., LXI (1959), 377–404, gives a clear review of anthropological trends in the United States, while Franz Boas, "The History of Anthropology," *Science*, n.s., XX (1904), 513–24, as well as his essay "Anthropology" in the *Encyclopaedia of the Social Sciences*, II, 73–110, and "The Limitation of the Comparative Method," *Science*, n.s., IV (1896), 901–9, offer critical analyses of early anthropological methods. Other good summaries include Alfred L. Kroeber, "History and Anthropological Thought," in William L. Thomas, Jr., ed., *Yearbook of Anthropology—1955* (Chicago, 1956); Paul Radin, "History of Ethnological Theories," *American Anthropologist*, n.s., XXXI (1929), 9–33; Edward B. Tylor, "American Aspects of Anthropology," *Popular Science Monthly*, XXVI (1884), 152–68; Robert H. Lowie, "Reminiscences of Anthropological Currents in America Half a Century Ago," *American Anthropologist*, n.s., LVIII (1956), 995–1016; Lucile E. Hoyme, "Physical Anthropology and Its Instruments," *Southwestern Journal of Anthropology*, IX (1953), 408–30. Aleš Hrdlička, *Physical Anthropology; Its Scope and Aims; Its History and Present Status in the United States* (Philadelphia, 1919), gives an excellent summary of physical anthropology in the nineteenth century. See also H. M. Hoenigswald, "On the History of the Comparative Method," *Anthropological Linguistics*, V (1963), 1–11; George W. Stocking, Jr., "French Anthropology in 1800," *Isis*, LV (1964), 134–51; and Katherine George, "The Civilized West Looks at Primitive Africa: 1400–1800. A Study in

Ethnocentrism," *Isis*, XLIX (1958), 62–72. There is also Leslie A. White, "Evolution in Cultural Anthropology: A Rejoinder," *American Anthropologist*, n.s., XLIX (1947), 400–413; Juan Comas, "Recent Research on Racial Relations," *International Social Science Journal*, XIII (1961), 271–99; and Stanley M. Garn, "Race and Evolution," *American Anthropologist*, n.s., LIX (1957), 218–23.

With respect to those works of the late nineteenth century which served as building blocks for the racial attitudes of the century, the reader should be aware of Benjamin Kidd, *Social Evolution* (New York, 1894); Thomas Bendyshe, ed., *The Anthropological Treatises of Johann Friedrich Blumenbach* (London, 1865); Armand de Quatrefages, *The Human Species* (New York, 1879) and *The Natural History of Man: A Course of Elementary Lectures* (New York, 1875); Lambert A. J. Quetelet, *A Treatise on Man and the Development of His Faculties* (Edinburgh, 1842); Augustus H. Keane, *Africa* (London, 1895), *Ethnology*, 2nd ed. rev. (Cambridge, 1896), and *Man, Past and Present* (Cambridge, 1899); Ernst H. Haeckel, *The Evolution of Man: A Popular Exposition of the Principal Points of Human Ontogeny and Phylogeny*, 2 vols. (London, 1879), and *The Last Link: Our Present Knowledge of the Descent of Man*, 2nd ed. (London, 1899); and Thomas H. Huxley, *Man's Place in Nature and Other Anthropological Essays* (New York, 1894). Paul Topinard, *Anthropology* (London, 1878), is a concise, thorough study of physical anthropology. His *L'anthropologie aux Etats-Unis* (Paris, 1893) gives a good perspective of American anthropological development. He begins with a general outline of anthropology and moves on through methods of race classification, the instruments of measurement, and the origination controversy. Another important source for American anthropological development is Otis T. Mason's yearly progress accounts in the Smithsonian Institution's *Annual Reports* from 1881 to 1888. He divided his reports into such sections as comparative psychology, phrenology, ethnology, sociology, and the instrumentalities. There is also John W. Jackson, *Ethnology and Phrenology, as an Aid to the Historian* (London, 1863); Thomas Bendyshe, "The History of Anthropology," Anthropological Society of London, *Memoirs*, I (1863–1864), 335–458; C. Staniland Wake, "The Adamites," Anthropological Institute of Great Britain and Ireland, *Journal*, I (1871–1872), 363–76; James Hunt, "On the Application of the Principle of Natural Selection to Anthropology," *Anthropological Review*, IV (1866), 320–40; and Benjamin A. Gould, *Investigations in the Military and Anthropological Statistics of American Soldiers* (New York, 1869).

For those works dealing with the interrelation of the sciences, the reader should become acquainted with John P. Gillin, ed., *For a Science of Social Man: Convergences in Anthropology, Psychology, and Sociology* (New York, 1954); and William F. Ogburn and Alexander Goldenweiser, eds., *The Social Sciences and Their Interrelations* (New York, 1927). Herbert Spencer, *The Classification of the Sciences: To Which Are Added Reasons for Dissenting from the Philosophy of M. Comte* (New York, 1864), explains the cosmic relationship of the disciplines to each other and the importance of the comparative method in bridging the concrete sciences. Then there are Maurice Greer Smith, "The Influence of Anthropology on Sociology," *American Anthropologist*, n.s., XXXI (1929), 819–22; Spencer F. Baldwin, "Present Position of Sociology," *Popular Science Monthly*, LVIII (1899), 811–22; Joseph LeConte, "Scientific Relation of Sociology to Biology," *Popular Science Monthly*, XIV (1879), 425–34; Franklin H. Giddings, "The Province of Sociology," American Academy of Political and Social Science, *Annals*, I (1890), 66–77; Robert H. Lowie, "Psychology and Sociology," *American Journal of Sociology*, XXI (1915), 177–269; and Albion W. Small, "Fifty Years of Sociology in the United States," *American Journal of Sociology*, Index to vols. I–LII (1895–1947), 532–54. Alfred L. Kroeber, "History and Science in Anthropology," *American Anthropologist*, n.s., XXXVII (1935), 539–69, offers a thoughtful summary of late nineteenth- and early twentieth-century thought concerning the concrete science series. See also Charles A. Ellwood, *Some Prolegomena to Social Psychology* (Chicago, 1901); G. Archdall Reid, "The Biological Foundations of Sociology," *American Journal of Sociology*, XI (1906), 532–54; John Wesley Powell, "Sociology, or the Science of Institutions," *American Anthropologist*, n.s., I (1899), 475–509 and 675–745; Carlos C. Closson, "A Critic of Anthropo-Sociology," *Journal of Political Economy*, VIII (1900), 397–410; Charles Hunt Page, *Class and American Sociology: From Ward to Ross* (New York, 1940); David G. Ritchie, *Darwinism and Politics* (London, 1889); Howard Becker and Harry E. Barnes, *Social Thought from Lore to Science*, 2nd ed., 2 vols. (Washington, D.C., 1952); Harry E. Barnes, ed., *An Introduction to the History of Sociology* (Chicago, 1948); William T. O'Connor, *Naturalism and the Pioneers of American Sociology* (Washington, D.C., 1942); Walter Bagehot, *Physics and Politics: Or, Thoughts on the Application of the*

Principles of "Natural Selection" and "Inheritance" to Political Society (New York, 1948); James P. Lichtenberger, *Development of Social Theory* (New York, 1938); Luther L. and Jessie Bernard, *Origins of American Sociology: The Social Science Movement in the United States* (New York, 1943); and Roscoe C. Hinkle, Jr., and Gisela J. Hinkle, *The Development of Modern Sociology, Its Nature and Growth in the United States* (New York, 1954).

There are also the superb studies by Kenneth Bock, "Darwin and Social Theory," *Philosophy of Science*, XXII (1955), 123–34; John C. Greene, "Biology and Social Theory in the 19th Century: Auguste Comte and Herbert Spencer," in Marshall Clagett, ed., *Critical Problems in the History of Science* (Madison, Wis., 1959); Gloria McConnaughey, "Darwin and Social Darwinism," *Osiris*, IX (1950), 397–412; John L. Myers, "The Influence of Anthropology on the Course of Political Science," University of California, *Publications in History*, IV (1916), 1–81; Lester Ward, "Neo-Darwinism and Neo-Lamarckism," Washington Biological Society, *Proceedings*, VI (1891), 11–71; L. H. Bailey, "Neo-Lamarckism and Neo-Darwinism," *American Naturalist*, XXVIII (1894), 661–78; Edward J. Pfeifer, "The Genesis of American Neo-Lamarckianism," *Isis*, LVI (1965), 156–67; and George W. Stocking, Jr., "Lamarckism in American Social Science: 1890–1915," *Journal of the History of Ideas*, XXIII (1962), 239–56.

Index